SCIENCE AND SENSE

Kelly Kading

ISBN EBOOK: 978-1-64873-531-8
ISBN Paperback: 978-1-64873-530-1

Printed in the United States of America
Published by: Writers Publishing House
Prescott, AZ 86301

Cover and Interior Design by
Writers Publishing House

Dedication

Science and Sense is dedicated to my estimable and beautiful wife, Jana. Her encouragement and input has been more valuable to me than she knows. Thirty-seven years of putting up with me is a testament to her grit and perseverance.

Foreword

Science and Sense is a rare combination of educational, accurate, concise and humorous observations - a great read for anyone looking for a balanced, unbiased take on today's environmental issues.

As a 30-year practicing environmental lawyer with an M.S degree, I've read papers, articles, memoranda, briefs and court opinions on just about every environmental topic. The political and agenda-driven, so-called "consensus science" arguments have increasingly infiltrated just about every environmental arena, to the detriment of actual science. Science and Sense shows that some scientists can still adhere to basic scientific principles in analyzing and arguing the environmental topics of the day, and explain their position in a rational, simple, relatable and entertaining way.

I've known Kelly for almost as long as I've been practicing law. We met in Phoenix in the late 1990s - I was a young environmental lawyer, and he a young geologist, and we travelled in the same environmental professional circles. We would discuss all sorts of environmental topics, with me providing a legal perspective and Kelly the geologist perspective. I was always impressed by his ability to make complex scientific issues understandable, and hopefully I was able to do the same for him on legal issues. We've remained friends to this day and continue to discuss the latest environmental issues whenever we get together. And I look forward to each column of Science and Sense not because I know the author but because I know the read will be fair, fact-based and fun!

~ Michael Ford, Esq. Attorney at Law.

Prescott Daily Courier's
"Rants and Raves"

- **Kudos to Kelly Kading's three-part column series** on climate, so educational and easily put. I felt like I was in an OLLI class that I have been reading about lately. I even did some research on my own since I felt challenged by him, very greatly appreciated!

- **Kudos to Courier Columnist Kelly Kading** for his climate (un)change columns. The truth will set you free!

- **Thank you, Kelly Kading**, for delivering the painful "climate change" truth to brain-washed fools.

-**Appreciate Courier Columnist Kelly Kading's** reasonable discussions of subjects too often too emotional.

- **Kelly Kading's columns** are always a joy to read and educational. I'm glad The Daily Courier didn't take the censorship advice from the sour grapes opponents who want him fired.

INTRODUCTION

In the mid 2010s, a high school friend (the estimable Marvin Longabaugh, lawyer and raconteur) had started a small-town newspaper in Navasota, Texas, and asked me to write an Op-Ed column about scientific issues, or whatever I wanted to cover. I had written for the high-school paper, and somehow Marvin remembered that I was a passable scribe. I wrote for the Navasota Star for just over a year, and enjoyed it immensely. Sadly, Marvin passed away in 2017 and the paper folded. The world is less fun without Marvin in it. See the Epilogue for more about my friend Marvin.

Time passed, and I was asked to write a similar column for the Prescott Daily Courier in early 2023. In the intervening years, "science" had become politicized to a sickening degree, and people were often duped by activists who touted themselves as "experts", when they had an axe to grind. Actual science had been subverted by "messaging", and it made my blood boil. My column, "Science and Sense" runs every other Sunday, and my Editor is gracious in giving me free rein, like Marvin did. I endeavor to provide the "straight poop", actual science, without political spin. I have managed to gain some fans, and also some detractors who just hate that my "common sense" often pokes holes in their (usually politically-driven) dogma. I rather like being the swizzle stick that stirs the Margarita.

Science And Sense is a compendium of some of those columns – the ones that I feel might resonate both locally and among people everywhere. Be forewarned - I do not use extensive bibliographies, because having to list sources would blow my limited space in the paper. Suffice it to say that anyone can look up on the internet the facts that I state, and find support (or counter-arguments) to my positions (these are opinion columns, after all). This is part of the goal of the columns – to make people think, research for themselves, and be more scientifically literate. My opinions are also based not just on what I have read and learned in 40+ years of scientific endeavor. I include my personal experiences working on projects in 34 states, traveling to all 50 states (and several other countries), and a solid dose of the West Texas common sense that was part of my upbringing.

The columns also try to limit the scientific jargon, and big words, as much as possible. In my career I have had to describe fairly complex geologic, chemical, and contaminant movement principles to engineers, clients and the public – therefore I have learned to "translate" these complex topics into terms that the average person can understand. So, don't expect to be dazzled by the minutiae and verbosity common in PhD treatments of these issues. My goal was always to write in a way that the typical newspaper reader can understand.

My hope is that you gain some insight into the topics presented, and are able to fight back against the "spin" so common in today's media. Doing your own research is also a goal of this book. The more informed you are about these issues, the less likely you are to be duped by the spin. Always remember, scientists are sometimes wrong, and they often agree with whoever is funding them. Go in with that understanding, and you are well ahead of the game!

Contents

CHAPTER I
General Science Topics

INTRODUCTION

This chapter begins with the very first column I wrote for the Courier. We have a local activist group here who presents themselves as "learned experts" on water and water availability, assuring that they just want to get out the facts, while ginning up fear that water is running out because of all the evil newcomers (mostly Californians). I was immediately suspicious, because their website states that they are anti-growth, and they ask for money for "memberships". Activist "science" and Actual science are two very different things, and the difference is "spin" for political gain (among the activists). I made no friends among this group, which is fine with me. I don't like it when people use their vaunted "expertise" to spin the opinions of those with less knowledge of the topic, for political reasons.

Next up is a brief introduction to the National Environmental Policy Act, for better or worse. Then there is a smattering of environmental topics – contamination scares, noise, and air quality. As an avowed weather geek, I couldn't help but include a column about the National Weather Service's

Weather Spotter program. The capper to this chapter is a treatise on "chemtrails", a litmus test topic for the conspiracy theorists among us. The column provides the actual, somewhat boring reality to rebut the "chemtrail people".

Enjoy!

Actual Science vs Activist Science

In my first column, I want to address the difference between Actual Science and Activist Science. In a world where the media SHOULD be calling out bad science, in most cases they choose a side and promote it. Activist science promotes a narrative that may or may not reflect the actual science. It is up to us to question our "experts," or resign ourselves to being useful sheep who blindly believe without questioning. And most people aren't scientists.

First, understand that actual scientists use the Scientific Method, where a position is proposed, then data is used to prove or refute the hypothesis. It is anticipated that someone will be checking the methods and making sure that all the assumptions are as neutral, accurate, and reproducible as possible. Good scientists always question, and there is no such thing as "settled science." If a hypothesis is proven false, it informs the next hypothesis, and the process continues until the truth is found.

Actual scientists are transparent in their methods, willing to let

anyone question their assumptions, and revisiting a hypothesis is normal, when new data arrives. Activist scientists are reluctant to divulge their methods, and are indignant when anyone has the temerity to challenge their dogma. Activist scientists also call out their dogs, and attack questioners or try to tear them down as "non-believers in science." The opposite is true – questioning IS good science. If an activist attacks questioners, then he/she is a poor scientist, and wants to hide his/her agenda. Activist groups employ lawyers, social media attack dogs, and any number of methods to silence opposition. A quick online search will show you that many activists groups work hard to shut down dissent.

Computer models (predictive models) are a great example of the difference between actual and activist science. I have used many predictive models in my career, usually contaminant movement models, often groundwater models. Other notable models are climatic models and disease spread models. But all models use variables – some known (like the force of gravity) and many unknown (different types of rocks' porosity, for example), where the scientist must "guess" within a range of possibilities.

A model can be manipulated one way or another by changing the assumptions in the variables. If you have five "range" variables, and you input the most conservative numbers for each, they have a multiplying effect to get to a conservative result. Run the same model with the five variables at the other end of the range, and the model result will be very different. Example – a contaminant might be shown by the model to be moving in groundwater at 5 ft/month, or at 500 ft/month, depending on how the modeler chooses the variables.

Climate models, for example, are notoriously bad at prediction. "Climate scientists" freely admit that their models have LOTS of variables, many for which our current data is insufficient to even create a range. So, they guess, and they often guess wrong. One climate model from about 1999 predicted that Florida would be underwater by 2009. Cocoa Beach (my last visit in 2019) is right where I first found it in 1967. Disease spread models are also notoriously inaccurate, especially with new diseases. Government agencies will always bias toward "safety" – we saw this recently with unnecessary shutdowns and other measures that did nothing, sometimes made things worse, and definitely had unforeseen consequences.

Groundwater models are also frequently inaccurate, basically because subsurface geology is not uniform, the data set is short (in terms of time), and we don't have enough data points (wells) in our county to fully understand the complexities

5

of the aquifers and sub-aquifers. But certain activist groups are quick to trot out models saying one thing, when the same model can come up with very different results, using other variable "guesses." And no matter how many members an organization has, or how many PhDs they have – if they have a position to push, their model results WILL push it. For example, pushing the narrative that groundwater supplies are dangerously low, to gin up fear and opposition to population growth.

The Arizona Department of Water Resources (ADWR), charged with protecting our groundwater resources AND assuring that groundwater is available for ALL uses (including human use), deals with groundwater models regularly. A developer's consultant might say "100 years of water, here's our model." An activist group might push back and say "Nuh-uh, we will run out of water next October, here's our model." ADWR scientists wade into the data (bad pun) and determine the best guess at accuracy, usually somewhere in the middle.

So, just like in the Wizard of Oz, watch the man behind the curtain closely. Be skeptical. Make the "expert" show their work, and especially ask them about variables that are "guesses" and how they dealt with that in the model. Ask how the sausage is made. Most importantly, don't just blindly accept a position from any "scientist," especially one with an activist position to push, or an axe to grind.

You might be surprised that your "expert" is really a salesperson, pushing a product (their position), not science. In other words, "Buyer (or thinker) beware."

The Science of NEPA

For those unfamiliar with the NEPA acronym, it stands for the National Environmental Policy Act. Signed into law on January 1, 1970 by the Environmental President (Richard Nixon), NEPA seeks to "encourage productive and enjoyable harmony between man and his environment..." among other things (quoted from the Preamble to the NEPA Act). The overarching goal of NEPA is to give the environment an equal voice to man regarding the construction of federally-funded projects. NEPA was partially a reaction to the frenzy of construction associated with the completion of the Interstate Highway System in the 1960s, which routed superhighways with little concern for the communities, habitats, and nature through which they were constructed. Also in the late 1960s a general increase in interest in the environment took root, evidenced by the raft of environmental laws passed in the early 1970s. Other issues were addressed by NEPA, but the bottom line is that the American people, and Congress, acted to broaden the factors considered when contemplating a federal project.

That sounds just lovely, logical and well-intentioned. In the early days of NEPA, environmental documents were straightforward and short. The first EIS I ever read was one completed in 1979 for a 70-mile highway widening project in New Mexico. It was 29 pages long, and had about 300 pages of appendices. Forty years later, the EIS for the 22-mile South Mountain Freeway project, Arizona's newest freeway (SR 202L, Maricopa County) was completed. I worked on this EIS off and on for 16 years of my career as the Hazardous Materials subject matter author. The final document was over 3000 pages, with several volumes of appendices (over 20,000 pages). What was the big difference in 40 years? In a word, lawsuits.

NEPA, and in California, the more stringent state Act (the California Environmental Quality Act, or CEQA) became a lawsuit machine for activist groups. Led by the Sierra Club, and joined later by every flavor of activist group you can imagine, they learned to delay or stop projects through what we now call "lawfare." Bogging down approvals in years of lawsuits added cost to the projects, and killed many. Were some of these lawsuits valid? Certainly. But most were used as a weapon against any number of projects that special interest groups opposed. The reason that the South Mountain Freeway EIS was so voluminous was that every technical topic was written in a way to fend off lawsuits. It was very effective, and the existence of this new, vital highway in Maricopa County is a testament to an EIS process and team that was laser focused and dedicated for over 16 years. This is true of all modern EIS documents –

they are no longer written to provide a succinct treatment of valid issues, they are fully "lawyered up," and of course, far more expensive to produce.

NEPA allows for two lower tiers of analysis, depending on how extensive the potential environmental impacts are for a stated project. I will use highway projects as my examples, but remember NEPA applies to any federally-funded project, even partially federally-funded. The first, and lowest, level of analysis is the Categorical Exclusion (CatEx). This level is appropriate for projects with very minor impacts – generally small additions to existing facilities, and takes a period of weeks to months to complete. The project is excluded from detailed analysis, because the critical categories (endangered species, wetlands, noise and air quality impacts, etc.) do not change from the pre-project condition. An example of a project where a CatEx analysis can be used would be a section of highway where the project is to add a broader shoulder, and upgrade the guard rails. All of the work would be constructed within the right-of-way, no additional land would be acquired, and the traffic volumes would not be affected.

The second, "middle" scope of NEPA analysis is called the Environmental Assessment, or EA. This level of analysis is appropriate if only SOME of the long list of environmental parameters might be changed by the project. The list of technical studies for an EA may be long or short, depending on the project. An EA typically takes months to a few years to complete. Endangered species analysis, wetlands inventory, hazardous materials issues, noise and air quality impacts, environmental justice, community cohesion, and others might be included in an EA. The specific issues to be addressed are part of the EA approval process. After all of the technical studies are completed, the only way the project can move forward without going to a full EIS is to get a "Finding Of No Significant Impact," or a FONSI. Aaaayyyyy Happy Days, right? An example of an EA project that is too complex for a CatEx, but doesn't rise to the level of an EIS, would be an expansion of a traffic interchange that would require additional right-of-way, property acquisitions, drainage changes, and an increase in traffic (noise and air impacts, for a start). The I-17/SR 69 interchange improvement a few years ago is an example of a project that went through as an EA.

Of course, projects that are located on virgin ground, or create major changes to the local landscape, usually goes to a full EIS (which takes many years to complete). The South Mountain Freeway is a good example, as well as the proposed Red Rock Rail Transfer Facility, being considered near Picacho Peak. And obviously, the new I-11 will go to a full EIS.

Next time we will go into detail on the many technical studies, and how they can be used to predict or mitigate environmental impacts from the built environment.

The Science of NEPA, Part 2
Technical Studies

Last time we covered the genesis and development of the National Environmental Policy Act (NEPA), and the different levels of study required for projects seeking federal funding. A number of technical studies underpin an Environmental Impact Statement (EIS) (the biggest document), an Environmental Assessment (EA) (for somewhat lesser impact projects), or a Categorical Exclusion document (CatEx) (least impactful projects). The EIS includes all the studies listed below, because on a big project, completeness is important (remember, special interest group lawsuits). For the EA, some of the listed studies are not performed, or are given limited attention (for example, no wetlands study if the project is located where there are no wetlands). For the CatEx, the document basically explains why the project will not cause impact to any category (the guardrail or shoulder-widening example from the last column).

Of course, all environmental documents set the stage with a statement of the "Purpose and Need" for the project – and there is more to this than you might think. "P&N" lawsuits are very popular. Introductory sections also spell out background information about the area and prior projects, listing of the cooperating agencies, coordination with local, state, and tribal entities (tribal consultation is a separate protocol), and alternatives considered and deleted (and why), plus the "No-Build" alternative. This preliminary information can take dozens of pages, but is important in setting the stage for the analysis.

The various studies fall into four broad categories (these are my own construct, to simplify explanation). These categories are Social/Societal, Earth Conditions, Human Considerations, and Special Studies. I have seen EIS and EA documents with all or some of the listed studies, but the order and emphasis of how they are presented varies, depending on the sponsoring agency's preferences, and the preferences of the consulting firm that prepares the document.

First up in the pantheon of studies are the Social/Societal studies. These include Land Use in the project area (farms, open land, industrial, residential, etc), Social Conditions (demographics of the people that live nearby), Environmental Justice (consideration of any "disadvantaged populations" in the area), Displacements and Relocations (moving people or businesses out of an area needed for the project, and associated costs), and Economic Impacts (how the project might change the economic conditions of an area). All of these studies are generally done by a "NEPA Generalist", someone that usually holds a degree in Environmental Science, and has experience in this specialty.

Next in my descriptive list are the "Earth Conditions" studies. These include Air Quality, Noise, Water Resources,

Floodplains, "Waters of the US", Wetlands, Geology and Soils, and biological resources. These studies can be critical to the document, since NEPA is all about balance between man's needs (the project) and how the project might affect nature in the area. Specialists perform these studies, from a wildlife biologist or an endangered plant specialist for the biological analysis, to a Noise expert, to an Air Quality analyst, and so forth.

My category of "Human Considerations" includes topics related to human history or human interaction with the environment. Cultural Resources is the largest category here, which includes consultation with Native American tribes and pueblos for their special historical relationship with certain places and landscapes. This section of an environmental document can be extensive here in the Southwest. Prime and Unique Farmlands is another topic of human interest, and worth considering and preserving. "Section 4f and 6f" properties include things like cemeteries, parks, historic churches, and historically significant set-aside lands that most people would not want to see disturbed, or are legally protected. These studies are performed by specialists, especially Cultural Resource consultation, which requires meeting certain state and tribal qualifications.

Finally, "Special Studies" are near and dear to this author's heart. These topics turn the tables to a degree, and consider how the offending planet might interfere with the project. Hazardous Materials studies (my livelihood) involves locating current or historical sites that might have residual contamination that could affect the project's construction. Energy, Temporary Construction Impacts, and Material Sources and Waste Material are also special studies that are important to consider, and include cost analyses. I include "Secondary and Cumulative Impacts" in the Special Studies category, mainly because "S&C" is a black art that I have seen go sideways into differing focuses.

So, as you can see, federally required studies can require a large team of specialists, lots of time to prepare, and high cost. They can be as simple as a CatEx that takes a few weeks and under a dozen people to prepare, or they can be like the South Mountain Freeway EIS. I worked on that monster for 16 years, included in a team of dozens of specialists, at a cost of, well, look it up if you must. The big-project EIS is not for the faint-of-heart, nor the light-of-wallet.

Perhaps now you have a window into why federally-funded projects are so expensive, and take so long. And you might understand why states and local governments accept federal funding with trepidation – lots of strings come with that money. Many states and municipalities find ways to self-fund projects to avoid the full NEPA process (although agencies like ADOT, for example, still do some level of environmental analysis on state-funded projects). And I am glad that they do. It is

15

important for us to value the ideals of NEPA, best stated in the NEPA Preamble – a "Harmony" between man and nature.

The Science of Contamination Scares

This column will cover some public scares about various contaminants, both real, overblown, and imagined, that we have weathered in the past. This background will be helpful when considering "new and emerging" contaminants, and determining whether the risks are real, or if it is just a ginned-up brouhaha.

Our modern life currently includes tens of thousands of different chemicals, used in every product you can imagine. Some make paint last longer, others keep food fresh, others keep bugs away – I could go on forever. The important point here is that of these tens of thousands of chemicals, only a fraction have gone through human health impact analysis (generally known as toxicology testing). Even for those that have gone through toxicology testing, human reaction to chemicals is never completely cause-and-effect certain. I used to teach a course in Toxicology, and the point I always tried to make was that the testing protocol uses rats or pigs or dogs or mice, not humans. Also each individual human has different dose/response reactions. A favorite illustration I used in the Toxicology class was based on alcohol (the example, not the presentation). Two shots of tequila are likely to have more of an impact on a 90 pound person, as opposed to a 260 pound person. And for most people, one tequila is a tonic, with a pleasant buzz. Six tequilas are likely a poison, that will make you sick. And 25 tequilas, well, you figure it out. Dose matters.

Public contamination scares have ranged from the real, to the overblown, to the total farce. Figuring out which is the rub. For example, DDT is an insecticide that was widely used, beginning in about 1940. Uses included everything from agricultural pest management to malaria control to delousing soldiers. In the 1950s/60s, it was determined that DDT thinned the shells of eagle eggs, and had other detrimental effects to birds. It was mostly banned, allowing some limited uses (DDT is still used in Africa and Asia for mosquito control to battle malaria).

Although I am old enough to remember running behind the city's mosquito-fogging truck in summer (DDT had a sweet smell), they had stopped the practice by the late 1960s. So, my older siblings chased the DDT trucks for longer than I did, which explains a lot.

Industrial chemicals also have bad health effects on humans, in uncontrolled exposures. I won't belabor Love Canal (Niagara Falls, NY area), the mother of all public health scares. In that case, the chemicals, exposures, and resultant diseases were very real, and the carelessness of their disposal is

18

shocking. The EPA correctly evacuated the area, bought out homes, and got people away from the mess. Just five years later, however, Times Beach happened. Times Beach, Missouri was a little hamlet with dirt roads. For years, the town hired a guy to spray oil to keep the road dust down. He got the oil from a local power company – using their spent transformer oil. This oil contained Dioxin, a nasty byproduct of overheated transformer oil. Although no health effects were reported by the local population that could be connected to this practice, the EPA (still dealing with the Love Canal debacle) took some soil samples, and decided that the town was just too toxic, and moved everyone out. But (a BIG but) – they had no established concentration for toxicity of Dioxin. They hastily declared an "interim action level" above which the soil was deemed "dangerous," using limited toxicity studies. Many millions of our tax dollars were spent relocating residents, buying and demolishing properties, and digging out soils to go to a landfill. Here's the kicker – a few years later, when the toxicology studies were completed, it was determined that the "interim action level" was way too restrictive, and the town need not have been evacuated and demolished at all.

"Alar" is the trade name for a coating that was placed on apples post-harvest, to minimize spoilage until point-of-sale. Alar washes off easily. Toxicology testing had been done by the FDA in the late 1960s, and Alar was approved. In the late 1980s, an activist group got ahold of some additional testing, which included extremely high doses of Alar, which did cause cancer in some mice. The activists went wild, demonizing Alar in the press, and causing a dip in apple sales because people were suddenly terrified of washing an apple. Here's the kicker – the amount of Alar that caused cancer in mice was like a human eating hundreds of (unwashed) apples a day for a couple of decades. Like the Rona, "global warming," and other examples of bad science, people's emotions go wild when they don't understand the facts.

Radon was a biggie for a while in the mid-1990s. Radon is a mildly radioactive gas that occurs naturally in some areas, often where granite exists near the surface. Since radioactivity is a "charged" word that means death to most folks, radon exposure hit all the papers. This author performed radon testing during dozens of property assessments in the 90s, and ONE TIME I found an elevated concentration, in an unventilated warehouse storeroom in Utah. Exposure to radon, in 99+% of occurrences, results in less radioactivity exposure than a dental x-ray, and the condition is easily mitigated by simple ventilation. Once the real (infinitesimal) risks from radon were better understood, and the realtors calmed down, the radon scare dissipated like trace radon gas when you open a door.

Many more contamination scares exist, and a quick internet search can entertain you for hours. The point here is that toxicity, dose, and response are what matters with exposure to any chemical. And when you don't know even one of those variables, that leaves room for emotion and fear to creep in.

The Science of Noise

Now that the election is over, we have all had it up to here with noise. But today's column is not about bombardment on media with political attack ads (which most people just mute). No, today the topic is noise in our environment in general – common sources of noise, ways to measure it, differences in noise perception among humans, and how to mitigate noise from common sources.

"Noise" falls into several categories when assessing it for human impacts. "Background" noise includes sound generated from human activities in general (like airplanes, traffic, air conditioners, and industrial activity), sound generated by nature (birds, insects, babbling streams, leaves rustling, and OMG cicadas!), and concentrated sources of sound (airports, freeways, mines, my neighbor's shrieking 8 year old daughter, heavy industry). Sound is measured in "decibels," from zero (inaudible) to about 140 (gunshots up close, jet engines at 100 yards, stereos cranked up to 11). The decibel scale is not linear, but logarithmic. This means that with every ten decibels, the intensity of the sound goes up by adding a zero. For example (and these values are approximate), a loud motorcycle going by is 100 decibels (near the pain threshold), while the sound of power tools (90 decibels) are 10 times less intense, an alarm clock going off (80 decibels) is 100 times less than the loud motorcycle, residential traffic (70 decibels) is 1000 times less than the loud motorcycle, and normal conversation (60 decibels) is 10,000 times less intense than the loud motorcycle. Damage to hearing can occur when regularly exposed to sound above 85 decibels.

Personally, I have assisted noise specialists with setting up and monitoring noise on major freeway projects in Phoenix. Every major roadway project includes noise analysis as part of the preliminary environmental studies. ADOT is a leader in the field of urban freeway noise abatement, evidenced by the hundreds (yes, hundreds) of miles of noise walls along Arizona's urban freeways. Noise walls block and divert freeway traffic noise up and over residential areas and human ears, mitigating the freeway noise to a large extent. ADOT also leads the nation in deployment of rubberized asphalt as a noise mitigation measure. On some Phoenix freeways, when you go from a concrete section to a rubberized asphalt section, inside your car it sounds like the engine suddenly turned off, a dramatic difference. That sound is also reduced out into the world.

Airports are a more difficult problem, since the noise source doesn't stay on the ground where walls will help. The best mitigation for airport noise is to not have residences near the airport flyways (off either end of active runways). The city of

Phoenix bought out hundreds of houses back in the 90s and 00s that were located off the east/west runways, and replaced them with less noise-sensitive warehouse-zoned areas. Here in the Yavaplex we have a busy and growing airport, and for some reason, residential properties have been permitted in the "noise zone." Our airport is a critical part of our economy, so "buyer beware" – the airport isn't moving, and more air traffic is inevitable. If you don't like aircraft noise,

you should consider buying a house that isn't near the airport.

Once the noise specialist has measured existing noise levels, they use computer models to determine how much more noise will be generated by a larger freeway, or at an airport, with more and different types of aircraft. Their product is a map that shows noise levels, where they occur, and what might be done to mitigate the noise increase. Interestingly, the way sound travels is not as straightforward as you might think. For example, the atmosphere has different layers, with different densities, which affect the way aircraft noise travels. For many years, my family lived 13 miles off the end of the runway at Luke Air Force Base in Glendale. The fighters would come over our house on approach to land, and overhead, they made a very unique sound (I like the sound of Freedom)! But they would get a couple of miles away, and it sounded like they shut the engines off for 20-30 seconds. Then the sound would come back. What was happening was a "bounce" of the sound waves off of a density layer boundary. The engines were never shut off, it is just that my ears were in a sound shadow as the angle of the noise bounced off the boundary.

Another complicating factor is that individual people perceive noise differently. Whether you have a sensitivity to certain sound bands, or if you are deaf in certain sound ranges, personal perception of sound varies. Remember that when a noise specialist prepares a model, they are using an adult with average hearing as the comparison baseline. My wife swears that I am deaf (when she talks to me in a normal voice from two rooms away). I have my hearing checked regularly, and I have no hearing impairments. Some people call it "selective hearing," but I will leave that domestic discussion for another column.

Bottom line, we live in a swirling, ever-changing cacophony of noise. My neighborhood on the edge of civilization is wonderfully silent after the sun goes down, quite a contrast to the constant noise I lived in during our Glendale years. It is amazing how you can get used to the constant buzz of air conditioners, urban traffic, and planes overhead. I still appreciate the silence at my place here in the Mayberry-in-the-Pines. I urge you to notice and think about the sounds you hear, both indoors and out. Bird song and the wind in the pines is a daily free concert. Listen up!

The Science of Air Quality

Here in the Mayberry-in-the-Pines, we don't think about air quality very much. Many of us came from Phoenix, southern California, or other urban areas where air quality is a daily concern. The one condition that makes us acutely aware of air quality, however, and it can be a huge deal here, is a nearby forest fire, when the smoke envelops us. The purpose of this article is to refresh us (or in some cases, introduce folks) to the fine art of deciphering air quality.

Generally, we worry about four categories of air pollutants that can make air quality a problem – ozone, PM10, PM2.5, and pollen. There are other things in the air that can mess with our lungs, that we don't get much of here (like volcanic ash, or heavy particulates from farming operations like plowing and spraying). If you want to go deep into air quality, a quick search of the "Innertubes of the YouGoogleWebs" will offer dozens of great articles about air quality. What you will learn, very quickly, is that big cities have many more air quality problems than places like the Yavaplex. Hooray for us! Clean

26

air might be among the reasons you moved here. But let's talk about a few of the categories of air quality insults that we do have to deal with.

First, ozone. Ozone is generated from normal atmospheric processes, mainly from the air interacting with the ground, with sunlight, and with humidity. Ozone is also generated in fairly large amounts by lightning, which welds oxygen molecules together in a flash. Ozone is a triple oxygen molecule, and our lungs have to deal with this odd molecule when we breathe it in. As air quality contaminants go, ozone is a minor player, not something we can do much about, and is just part of nature.

Second, PM10 and PM2.5 particles. This is basically dust in the air, with PM10 being a sort of average suspended dust particle size, and PM2.5 being ¼ the size of big brother PM10. Breathing in these dust particles is also just part of living on this dusty planet, however the smaller particles can penetrate deeper into the lungs. Normal lung function will bring up these particles to be expelled, but if you have a lung function problem such as COPD, asthma, or emphysema, your body has to work harder to expel this material (or you may not be able to expel it at all). In Phoenix, PM10 and PM2.5 particles exist at a higher concentration than in, say Boston, because Phoenix is located in a windy desert. The LA area has lots of particulates in the air because of high levels of traffic, exacerbated by the marine humidity which gives particles something to ride on. If you research areas on the planet with high particulates, you will find that desert areas, windy areas, or high population areas almost always have the highest counts.

Pollen and biological elements in the air can be a HUGE problem for some people. This author, for example, has a serious problem with mountain cedar pollen. Every spring, my sinuses explode because my body cannot deal with that specific kind of pollen (an allergy). Fortunately, I have found the magic bullet of a Kenalog shot (a steroid), which eliminates 90% of my suffering. Everyone is different, though – cottonwood trees, ragweed, animal dander, mold – there are a million different irritants in the air. Best to figure out which specific one bothers you, and come up with a plan of attack, be it over-the-counter meds, a Kenalog shot, or going to Alaska or the Bahamas for the spring.

Another category that is technically particulates, but can also include chemical irritants, are industrial chemicals in the air. I have friends and relatives in the Houston area, where the air is a toxic soup of big-city contaminants, refinery waste products, and a

potpourri of emanations from port facilities. I used to leave Houston with an upper respiratory infection almost every time I visited, but I am thrilled to report that I never have to go to Houston again for the rest of my life! LA, San Francisco, Seattle, Chicago, NYC, and Baltimore, however, are almost as bad as Houston, so be aware. And cities in India are hundreds of times worse than LA, mainly because they burn animal dung for cooking and heat, and they burn trash in the street, leading to visibilities of under ¼ mile in the cities. My trips to India have been followed by upper respiratory infections. Rural India is much cleaner, though, as is true almost everywhere (rural beats urban).

Finally, a few words about our sleeping air quality monster here in the Yavaplex. A forest fire, even a small one, vaults hundreds of kinds of particulates, pollens, combustion byproducts, and biological nasties into the air. I am sure that most of us can remember a smoke event in recent years, and it really doesn't take much to set your eyes to watering and your lungs convulsing in coughs, trying to expel the airborne insults. We have two great methods to limit our exposure to forest fire smoke – first, not have the fire (we can only affect that risk in a small way), or second, bug out when the smoke approaches, especially if you have sensitive lungs or a medical condition.

Meantime, count your lucky (and visible!) stars that you live where you do, and enjoy breathing our generally clean, fresh air. And if you really want to experience clean, fresh air, visit Barrow, Alaska – which regularly ranks among the cleanest air on the planet. Just don't freeze to death.

Allergens (ah-CHOO)

As a geologist, my understanding of botany is limited. Plants (and soil, for that matter) are those things that get in the way of easy observation of rocks, structures, and formations. But my time on the planet (and working closely with biologists on environmental clearance projects) has taught me a fair amount about plants. I pay special attention to plants that stick, sting, or stink (we have LOTS of those in Arizona), or even worse, plants that launch a chemical attack on our bodies. The mode of attack is by pollen, spores, sap, excretions, or many other sorts of botanical weapons. My beloved dogs also launch allergens my way, as do some foods that I otherwise love. So, the following is my treatise on allergens, from the sufferer's standpoint, since I am at peak misery these days.

What is an allergen? It is any substance that elicits an unpleasant or unwanted reaction in a living body. Common examples are dog, cat or horse dander (yes, they are different), pollen (from grasses, trees, flowers or other botanical sources), and certain natural or enhanced chemicals in our food. The response to an allergen varies by your individual sensitivity, and the mode of response.

Since your response to an allergen is what you care most about, let's talk about the three most common allergic responses - eyes/ears/nose/throat, skin, and digestive system.

First, eyes/ears/nose/throat reactions. Most of us experience some level of itchy/red eyes, nose, throat, and even ears in the spring. Pollen is plentiful at this time, and pollen reacts with your body two ways – chemically and physically. A chemical reaction occurs in, say, your sinuses, where your body goes into riot mode in response to a pollen to which you are allergic. Your body releases histamines to fight off the insult, resulting in runny nose, watery eyes, and itching where you can't scratch. The physical aspect of a pollen is revealed at the microscopic level. For example, my bugaboo, mountain cedar pollen, is not only a chemical problem for me (I am allergic to it), but the pollen itself looks like a spiky weapon of battle in a microscope. Ragweed pollen is another common pollen with a medieval shape, designed to inflict misery.

Second, skin reactions. Your body's shell (skin) can react in several ways to an allergen. Rashes, hives, and sensitivity are the most common. Penicillin, a mold and also a common antibiotic, is an allergen to me. The last time I had it, I broke out in itchy welts all over, my first experience with hives. An equestrian friend breaks out in a rash if exposed to horses (but not dogs or cats). This skin irritation is, again, your body releasing histamines in reaction to contact with an allergen.

Poison ivy or poison oak are different in that they are not technically allergens, but chemical irritants.

Third, digestive system reactions. Your body can react internally to an allergen as well. Most commonly with foods (but also with pollens that you inadvertently ingest), your digestive system can react in many unpleasant ways. Stomach aches, sharp pains, cramping, dull ache, loss of appetite, and an (ahem) urgent need to stay very close to the bathroom. Just like any allergic reaction, your body goes into battle mode, releasing histamines which wreak havoc with the delicate balance of your digestive flora.

Wow, these all sound horrible! you say. Well yes, they are. But you can battle these awful symptoms in different ways. If you know you have an allergy, and can avoid the allergen, do so. I am very allergic to cats (mildly dogs), so I stay out of places where cats live (our friends with cats just meet us at a restaurant). For pets, regular baths do wonders to limit dander. I know that cayenne pepper (but not other peppers) play hell with my digestive system, so I avoid the blackened fish ("blackened" usually means lots of cayenne). Avoidance works, when it can be done. You can also take a series of inoculations that are designed to de-sensitize you to your specific allergens.

Pollens (and mold spores) are something you can't avoid. They are in the air, largely seasonal, and prolific (think about the yellow dust on your car every morning in the spring). Well, you can avoid it, but it means using your car or an airplane to go very far away, to a place without the pollens that bother you for a few months (perhaps Antarctica). But that's not practical. No, you have to do battle with some allergens that you can't avoid.

Over-the-counter or prescription medications can limit your body's response to allergens. I know, everyone is trying to get away from another pill or shot. But given the choice between misery for three months, or medication, I know which way I am going. Pills (and diet and avoidance) can go a long way to restoring quality of life for the short period that you are affected. I have discovered the magic bullet (for me), is a steroid called Kenalog, administered as a shot, that lasts about two months (for me). I schedule it every February (this year I might need a second shot in May).

So to wrap up, know your personal poisons, practice avoidance when you can, and do chemical battle with them to maintain quality of life. We all love flowers, and our pets, and fresh food from the garden, so buck up. Allergens are a price we pay to live on this beautiful blue marble in space.

The Science of Weather Spotting

Most people have at least a passing interest in the weather. If you grew up on a farm or ranch, the weather forecast was vital. If you enjoy outdoor activities (hunting, fishing, hiking, boating, camping, rockhounding), it is important to know generally what the weather is doing. As for me, I grew up in flat, boring west Texas. West Texans are very much in tune with the sky, because the landscape is so flat and dull, and the sky offers some spectacular variety. Tornadoes, dust storms, endless droughts, sudden downpours, supercell thunderheads you can see for 80 miles, arctic blasts – I grew up in a land of weather extremes.

My interest went a step further in college. Studying Geology at Colorado State, I was generally looking down at the rocks, but occasionally I would stretch my back and neck, and look up. It occurred to me that perhaps I should learn about what was over my head, as well as under my feet. I mean, I had a whole university at my disposal (a leading school for Meteorological study), and some slots available for electives. While doing a Geology project at CSU's research flume (a controlled laboratory mimicking a miniature river), I started talking to the weather geeks who used the atmospheric instruments in the next building over. Next thing I knew, I was taking Meteorology classes. Everyone should be so lucky!

I have been fortunate enough to experience tropical rain in Miami (a totally drenching experience), 4 feet of snow in a day and -24 degrees (Minnesota), 117 degrees and a total blackout haboob in my yard in Glendale, strobe lightning (Indianapolis), thundersnow, two lightning strikes within 100 yards, softball-sized hail (three times!), one tropical storm, and I have personally seen seven tornadoes on the ground. These are the trophies on my weather geek mantel.

As a geologist, looking down at the rocks, I have received whiplash by keeping my head on a swivel to the sky. Both our Earth and the sky offer incredible beauty, as well as scientific and artistic endeavor. I developed a photography hobby which follows me to this day, but the photo image is a poor reproduction of being out in the middle of it - 3D beats 2D every time. I went along like this for most of my adult life, then later in life I found that I could take this personal interest and turn it into an altruistic endeavor. I became a National Weather Service (NWS) Skywarn Weather Spotter. Okay, don't get too impressed (autographs are $10).

With the inevitable approach of the Monsoon season, this is a great time to cover the Skywarn Weather Spotter program. Anyone can become a Spotter, by taking a simple two-hour NWW Spotter course. For me, the class was 95% review, since I was already well-familiar with cloud types and methods of observation and measurement. But I did learn the NWS process for making a report, what they need to know, and how to be a useful Spotter.

Here is what you'll learn in the NWS Skywarn Spotter class:
- Basics of thunderstorm development
- Fundamentals of storm structure
- Identifying potential severe weather features
- Information to report, and how to report
- Basic severe weather safety

I have an electronic weather station in my back yard. They can be bought for $50 up to several thousand, mine was about $300 and has almost everything I need, except a good snow trap (Christmas present HINT HINT?). But you don't need to have any special equipment to become a valuable Spotter – just your eyes, a cheap rain gauge, and the knowledge to understand what you're seeing.

The NWS takes in reports from Spotters, and uses that information to fine-tune their forecasts and warnings (in extreme weather). NWS explains that in Flagstaff they have a great array of state-of-the-art sensing instruments, but our mountainous terrain "blinds" the instruments in some cases. Here in Prescott, for example, what is happening down in the valleys is not really visible on NWS radar out of Flagstaff or (especially) Phoenix. Spotter reports provide the "granularity" (their word - forgive them, as they are science geeks like me) to fine tune a forecast or warning. And, you can be as involved as you like. I try to report when we have something unusual, but certainly not a daily report of mundane conditions.

You, the Spotter, also get something out of it. You get to learn more about our complex atmospheric phenomena, and you get a better understanding of our world. You get to learn some new words to impress your friends (my favorite is 'graupel," the little pea-sized snow pellets that we got a lot of this winter and spring). And you get that warm feeling inside by doing something altruistic – using your skills to help your fellow man.

I hope you take me up on the challenge, and also become a Spotter. Maybe we can get together over a beer sometime, and geek out over weather stories!

The Science of Lines in the Sky

Recently I was drawn into a social media conversation that astounded me. Someone had posted a picture of the sky over our wonderful town, on the day before Thanksgiving. Many white contrails (short for CONdensation Trails) were oriented east/west, toward the LA megaplex. Not coincidentally, a major flight corridor exists over Prescott, and on any given day you can see jets and condensation trails heading to and from the many airports in Southern California. The photograph in question showed a dozen or so. Kind of a neat photograph, so I waded in and started reading the comments.

I was gobsmacked to see that the majority of people were convinced that the contrails were, in fact, something they call "chemtrails". These people believe that the government (or big Pharma, or the Chinese, or LeBron James, pick your villain) is poisoning us from the sky, by spraying chemicals into the atmosphere. Some thought that cancer, Alzheimers, obesity, hair loss, societal anger, halitosis, or stupidity were caused by this nefarious aerial spraying program. Some people reposted articles from laughably unreliable sources, as their "proof" that "chemtrails" are real.

After a few moments absorbing that these people live among us, I crafted a response that I will summarize in the following paragraphs. Some of the "chemtrail" devotees scurried away in the face of actual scientific facts, but some doubled down, eventually calling me "blind to the reality of 'chemtrails'"". Some people are un-teachable.

So, let's explore their beliefs one at a time.

1. Apparently to the "chemtrail" followers, all military, commercial, and private planes have tanks on board that spray every second they are in the air. All right then, where are these tanks located within a plane (specifically a commercial plane?) The entire interior of a jetliner is jam packed with passengers (REALLY jam-packed), luggage, cargo, fuel, control systems, water and sewage, wheels, peanuts, and expensive food. Where would these tanks go? And why, in the 60+ years that these people claim that this has been going on, has NOBODY ever photographed a plane being loaded with chemicals? Tens of thousands of pilots and hundreds of thousands of aircraft technicians would have to be in on the conspiracy, and have maintained 100% silence over 60+ years.

2. Who pays for these chemicals, and who pays to transport them all over the country (actually the world) to every airport? Imagine the cost of thousands of gallons PER FLIGHT being loaded with expensive death poisons. Not even the federal government (our tax dollars) has enough money to fund that.

3. Why do the "chemtrails" exist wherever the jets fly, and only during certain atmospheric conditions? Seems like an awful waste of money and chemicals to spray disease-causing poisons over the ocean, northern Nevada, most of Nebraska, or northwest Yavapai County if there are no humans to receive the dose. If "chemtrails" were real, it would have to be the most inept and misguided government program ever, and that is saying a lot.

4. If this really is a nefarious US government program, why do jets have white contrails behind them all over the world? What is the benefit to the US government in spraying the southern Pacific Ocean? Why do jets from every country have white contrails behind them? Are Lufthansa, Air Vietnam, Qantas, and Air Zimbabwe in on it, too? And what about private jets, like the ones John Kerry and Taylor Swift use every day? Why would private jets be in on the program? Who is paying them, loading the chemicals, outfitting their planes with expensive tanks and sprayers? Who tells them where to fly to spray the poison? Something tells me that Taylor Swift is not going to divert her private jet flight from New York to Kansas City to spray her assigned coordinates in Montana.

The biggest question of all about the "chemtrail" myth – why do people believe it? Is it easier to believe an unprovable fairy tale than to understand actual science? Well, perhaps –

since science requires employing actual knowledge and logic (science and sense) instead of making up schemes out of whole cloth. Hollywood stories are plentiful, and perhaps have clouded the minds of some people. A well-spun yarn sometimes becomes reality to the weak mind. News flash folks, ET was not real. The movie JFK was just a movie. This intellectual laziness has been proven out by the Rona insanity, Bigfoot, and "man-made climate change", so maybe a certain percentage of the humans are gullible enough to buy this tripe.

So, here is the truth, "if you can handle it" (thanks Col. Jessup). The white trails behind jet engines are condensation trails (contrails). When atmospheric moisture, present at all altitudes, passes through a hot jet engine, the moisture briefly melts, then refreezes as larger crystals, visible to the eye as a sort of linear cloud. It is just heat acting on moisture, folks. Clouds are formed by a similar process (heat from the sun, making more diffused clouds). Sometimes jets make no contrail, and that is when the air at their altitude is extremely dry. Sometimes the lines in the sky linger, because there is little wind that day at that altitude to disperse the wispy contrail. And yes, there are chemicals in the vapor – tiny amounts of jet exhaust elements.

And if you want to be technical, pure water vapor is also a chemical.

Okay, now run and panic over the fact that the government is trying poison us with di-hydrogen oxide. Also known as "water".

CHAPTER 2
Environmental Topics

INTRODUCTION

Having spent over 35 years in the environmental field, this chapter is my wheelhouse. After starting off with a column about the early, and effective, environmental regulations passed in the 70s and 80s, I go right for the topic that I was goaded into since my first column, the land mine of "Climate". This three-part series made readers of a "certain bent" absolutely lose their minds, because the series pokes holes in the weak logic that underpins the modern climate religion. The paper got a ton of comments from readers on this series, and surprisingly, most were positive. I take the topic one step further after the three-part main storyline, by delving into how universities further this fantasy and waste student's time and careers on "climate".

Next up is another three-part series, about the Superfund program which cleans up (sometimes) abandoned hazardous waste sites. While the EPA has gone wildly beyond their original scope, cleaning up abandoned hazardous waste sites was their original mission, and one that they have done a fairly good job at for several decades now.

This chapter tops off with an olive branch to whatever "bent" you have, by covering common-sense things that anyone can do to improve the planet, even in small ways.

If you are infuriated by the Climate series, I urge you to set aside your burning anger, and keep reading. The rest of the chapters are far less polarizing.

Environmental Regulations
Good, Bad, Ugly?

My first column two weeks ago raised the ire of some people, who think their voices are the only ones that matter. This column will regularly provide non-activist information, to inform the non-scientist, average person on the basic science on topics of the day. Lest the fine readers of this paper think I am some anti-environment nut, allow me to disabuse you of that suspicion with this column.

Although my positions, rooted in **both** science and common sense, will surely ruffle feathers on either side of any issue, rest assured that I will always strive for balance, based on what we know, recognizing what we don't know, and suggesting what is the best scientific compromise to help mankind. One thing I can heartily endorse, and is the focus of today's column, is the general effectiveness of our early environmental laws. My job for 35 years is/was (I am semi-retired) to assess and clean up hazardous waste sites. I have seen firsthand what we have done to Mother Earth, and how she often just shrugs off damage through natural buffering processes. It is pretty inspiring.

I am old enough to remember the late 60s / early 70s, when air and water quality was horrible (even dangerous in cities like Houston and LA). I can remember rivers actually being *on fire* because of pollution (Houston Ship Channel and the Cuyahoga River in Ohio, to cite just two examples). Our nation before environmental protective laws was a free-for-all for industrial pollution, and we all suffered for it.

But beginning in the early 1970s, President Nixon (surprise!) signed the first few environmental protection laws – the National Environmental Policy Act, the Clean Air Act, the Clean Water Act, and several others. The effect of these laws (and the specific regulations that followed) had an amazing effect over time. Polluters (generally industrial, oil, and chemical companies) led the way in "cleaning up their act" so to speak, at least to a point that they wouldn't be shut down or fined by the EPA. Auto manufacturers were forced to address tailpipe emissions by the Clean Air Act, and even though early efforts robbed vehicles of our beloved horsepower, over time the auto industry figured out how to address emissions while preserving driveability and power. To wit, our current cars are far more powerful and hundreds of times cleaner than their 1968 versions. Mom's family sedan today provides similar horsepower to early "muscle cars", with a tiny fraction of the tailpipe emissions. This is why I can see the San Gabriel mountains when I visit LA, when in 1971 I couldn't (also because my eyes and nose were burning from the polluted air).

The cumulative effect on our environment has been miraculous. We live in a highly industrialized country (although a lot of manufacturing has moved to other countries, due to ridiculous taxation/over-regulation, but that's another topic). We have more cars per capita than any other major country (China has more in total), and our many cars **do not** cause choking air problems. We have five to ten times the cars now that we did in those cities in 1968, but today's cars emit a tiny fraction of the contaminants of a 1968 car. Do the math – this is why we can now breathe (most days) in big US cities, although we do a lot of that breathing while stuck in traffic. Congestion is a topic for another column.

For the past 25 years or so, the EPA has strayed from its original mission, and been weaponized for political pursuits. In fact, their extremism has done less to help the environment, and more to unnecessarily restrict business. The EPA has gotten involved in many areas where they don't even have expertise. Climate? We have the National Weather Service, National Center for Atmospheric Research, and another ten or so agencies that are directly or peripherally charged with climate/weather tasks. Even NASA – why in the world is a space agency wasting time on "climate"? (answer – funding). The EPA should stick to their core mission, cleaning up hazardous waste sites.

Another factor is all these "environmental science" (LOL) degrees that woke universities are handing out, usually with more political science curricula than hard science. These young graduates want to do something that they feel is impactful (as they have been told their degrees and input will support). But what is difficult for them to fathom is that the big pollution issues were handled decades ago, before they were born. If students really want to get involved in helping the planet, get a degree in the hard sciences (Chemistry, Geology, Biology, etc). Environmental Science degrees only prepare the workforce for activism, not useful careers. If they really want to make a difference, take that hard science degree and get an environmental cleanup job in China, or India (where I have visited). Those countries *really need* those skills.

So, the next time some greenie weenie (likely some kid under 50) gets all emotional about how we are "ruining the planet", just show them some archival photos of the Houston Ship Channel, or LA, and compare them to today. The laws that Richard Nixon (the Environmental President) signed, have *worked*. For comparison, go to India and see what happens without enforcement of environmental regulations.

The end result of these regulations is our amazingly clean country, and regardless of the politics, our early environmental regulations seem to work pretty well. And that is a testament to us as a country and a people, and what we demanded. We demanded cleaner rivers and air, and we got them. Government is an ugly bicycle, but like it or not, we learn to ride it, and sometimes it gets us where we want to go.

Climate Part 1

Ever since my first column, readers of a certain bent have been trying to goad me into the "Climate Change" fight. Normally I don't cover things like voodoo, chemtrails, or Bigfoot, but you insisted. I am sure it is some kind of litmus test for some of you. Whatever. This article is the first of a three part series, because the subject is a bit too complex for one column. I fully expect pushback for whatever I say, since the "Climate Cabal" is much more like "political science" than actual science, and people get very passionate about their politics these days. So, here we go – gird your loins. As we have learned from modern politics, armor is crucial.

As a Geologist, of course I believe in a changing climate. The climate has been changing since our planet was born. Geologists study the rock record, which goes back billions of years, with demonstrable evidence of the climate through depositional characteristics, plants and animals, chemical capture, etc. Geologists see that our planet has experienced much larger and faster temperature changes than what we are currently seeing, and it all happened before the modern industrial revolution, which accelerated the burning of petroleum and coal. In fact, natural processes drive climate much more than anything Man does, but I will come back to that later.

Warmists (the Climate Cabal) focus on the last 125 years or so, since the beginning of the industrial revolution, with the goal of blaming "climate change" on man's activities. Remember, 125 years, compared to the scale of geologic time, is 1/1000th of a blink. Why do they want to blame man? Power and control (more in a later column).

The following illustration of the folly of the warmists' short-sighted focus is apropos. Between November and January each year, I gain about 15 pounds from all the goodies, delicious holiday meals, and slowed-down outdoor activity. The warmist would look at that six week period, extrapolate it for five years, and conclude that in six years I will weigh over 700 pounds, and likely be dead. What the warmist fails to consider is the other 46 weeks of the year. In January-March I strive to eat better and work out more. Then April comes, and I work in the yard more, walk the dogs more often, and generally increase my physical activity. By May, I am back to my fighting weight (such as it is for a guy in his 60s).

This is the warmists' scam. They take little bits of cherry-picked data, and extrapolate it as if other conditions will remain constant, which they don't. Climate changes, it fluctuates. The planet "breathes", in natural cycles. Just one example is the El Nino / La Nina effect, which affects regional climate in North America on a scale of a few years at a time. El Nino cycles (typically 1-4 years) include a warming of ocean waters in the eastern Pacific (La Nina is cooling of those waters, similar timeframe). The last few, very wet months here in Arizona have shown that we are experiencing another shift, all CLIMATIC FLUCTUATIONS WITHIN A NORMAL RANGE. Activist "scientists" look at such a short data set, and blame it on one cause - "Man's activities". Even the vaunted "hockey stick graph" which warmists use to try and prove their theories, doesn't work anymore. That brief warming jolt ended about 20 years ago, when global temperatures started coming down (normal fluctuations). It is also interesting to look at 100 years of fluctuating solar output vs Earth's surface temperature. It is a near-perfect correlation – higher solar output years = warmer Earth years. A rather inconvenient truth, wouldn't you say?

The bottom line, in both the geologic record and in observable conditions, is that CLIMATE FLUCTUATES. Some years the changes are more sharply felt than other years, but over the past 125 years (and the past 125,000 years), these FLUCTUATIONS ARE WITHIN A NORMAL RANGE. Some years, we have lots of hurricanes and tornadoes. Other years, less. What is laughable to me is the warmists' claims of "record warmth!", "record cold!!" "record records!!!" (for records only kept for about 125 years, which is nothing). All breathlessly reported along with "see, we're killing the planet" proclamations. The key point is that 125 years is what we actual scientists call an "insufficient data set". Climate takes hundreds to thousands of years to significantly change (the last Ice Age was a mere 10,000 years ago, yet another blink in geologic time, and we are still warming from that event). So, the Climate Cabal's Chicken Littles run around saying "See! The sky is falling! A big hurricane! Your evil oil-burning Subaru caused it!" Throwing guilt is one of their favorite tactics. And the press jumps in with their hype, because "normal fluctuations" doesn't sell papers, or support the dogma, like "Climate apocalypse!!!"

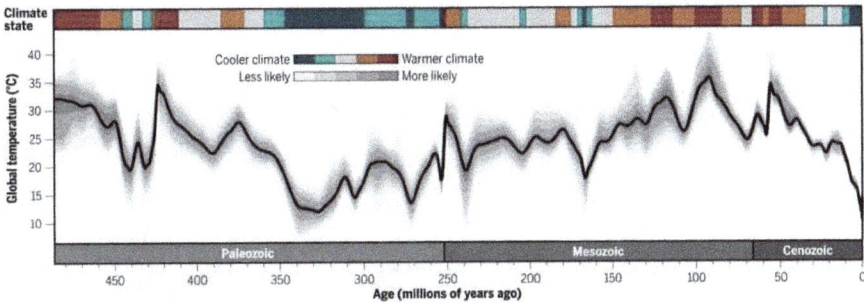

(this is average Earth temperatures, gleaned from geologic data, for the past 500 million years)

The warmists' touchstones – the IPCC (the politically-motivated International Panel for Climate Change), the "hockey stick graph" (going the other way now), Algore's soothsayer predictions (always wrong), "97% of scientist agree" (they don't), all pale in comparison to the 4 billion year rock record, which, if one takes time to study it, makes the warmists' histrionics all the more unsupportable.

I fully expect an absolute conniption from the warmists, even for this first article. Remember it is a three-part series, and each article will take on a different part of this complicated issue.

Keep your eyes peeled for their rebuttals, which will be filled with "peer-reviewed" paper (LOL) references, indignance about anyone having the temerity to question their conclusions, and personal attacks on me (attacking opposition is what warmists do, like a religious war). They will call me a "climate denier", but as I said earlier, I fully recognize that the climate is ever-changing. I urge you to read all three columns before you light your fuses.

Climate Part 2

Well, the conniption that I forecast in my first Climate column is in full swing. The academia/government axis is lobbing rhetorical grenades, accusing me of not being a scientist, being stupid, not falling in line, and generally being a terrible person. People who know me would disagree, but the personal attacks slip my shoulders like I'm made of Teflon (oops, a useful petroleum-based chemical). In this article, Part 2, allow me to explain why they are so angry.

Academia and government are like any endeavor – those engaged in it must do certain things to get ahead in their careers. I don't mention the private sector, because their motivations are obvious – doing good work for clients and customers, and getting rewarded by repeat business and profits. A chunk of those profits fund government (through taxes) and academia (through exorbitant "tuition and fees"). Government types get ahead by "solving problems" (they don't really do that very much; they APPEAR to solve problems). Often they create the problem, or gin it up, to get votes. Once they have voters whipped up that Problem A is going to ruin their lives, they need pseudo-solutions and gravitas to prove that their solution is the one. So, they provide grants (your tax money) to academia, so that professors can say "harrumph, harrumph, yes Senator Smith's solution will fix the problem".

Grant money (sometimes called "research" as a cover) flows through the university to grad students and professors. How do "climate" academics get ahead? Well, they write "peer-reviewed" papers, preferably about something that can't be proven, and their assumptions are pulled from (ahem) thin air. Man-made Climate Change (MMCC) is perfect, because you can't prove anything, or you can prove everything by claiming that normal weather is some indication of nefarious MMCC, caused by you driving your Subaru (for shame!) To be clear, universities do a lot of legitimate research, which is needed and appreciated. But climate research is much closer to witch-doctory than actual science.

As an aside, I have several meteorology and climate research friends. They have differing opinions, but they all report that they don't dare publish anything that challenges MMCC, lest their careers suddenly end. We like to think that tweedy professors and ivy-covered universities are genteel places, but they aren't. They defend their turf (dogma and funding) as viciously as any 1930s Chicago gangster.

The "peer-review" process, which is the glue that holds together the fake gravitas, is an exercise in mutual back-scratching. Academician A gets his unsupportable paper "reviewed" by Academician B, who needs his unsupportable paper peer-reviewed. So he goes to Academician C, who calls on Academician A to review his unsupportable paper. This is the "merry-go-round of baloney" that allows an unprovable hypothesis like MMCC to get some traction (among those who don't employ independent thought).

When it comes to MMCC, nobody ever talks about natural atmospheric and oceanic buffering. 71% of Earth's surface is ocean, and the atmosphere is a giant moving scrubber whose chemical reactions in the air (and with the ocean) destroy or remove contaminants. The total of man's annual input to the ginormous worldwide airshed is but a pinprick, easily managed by the atmosphere and oceans. So the warmists have decided to call CO_2 a "contaminant", to get the volume of "contaminants" that they need for their manipulated models. Sorry, atmospheric CO_2 is natural, necessary, makes up 0.04% of the atmosphere, and has been around since Day 1. We exhale CO_2, plants absorb it. Humans are MADE of carbon. Volcanoes can spew the equivalent of decades worth of man-produced CO_2 with one big eruption. The Earth's atmospheric and oceanic scrubbers deal with those inputs too. This is why after billions of years of eruptions, animal and plant activity, and 125 years of man burning hydrocarbons, we still have blue skies, clean water, and really no demonstrable effect on our world, despite whatever bad names the warmists call CO_2. Mother Nature is powerful, and her self-healing powers inspire awe.

Granted, man's insults to the air and water certainly can cause localized problems. If you live within a few miles of a coal burning plant or abandoned mine, you may have problems. If you live in a city (where most warmists live), you are surrounded by contamination sources, so the warmist thinks the whole world is like that. I urge urban warmists to stop screaming for a while, take a "sabbatical", and drive across northern Nevada. Or take a train trip across Russia (the long way, 9 time zones). Or visit the African veldt. Or take an ocean voyage from LA to New Zealand. Surprise! The urbanized places you inhabit are less than 1% of the Earth's surface. The other 99% is there, out of sight, quietly scrubbing out contaminants, or more likely never really being touched by them. Take off your urban, fear-driven blinders, and try to see things in a larger (global) context.

Of course, the traveling warmist on a voyage of discovery needs to take lots of pictures, and write articles about how the Earth is really okay, and the CLIMATE IS FLUCTUATING WITHIN NORMAL LIMITS. Those articles might calm down the children and non-scientific citizens that you have frightened into a lather over a phantom boogeyman. But remember, you won't get any grant money for saying "everything is alright". And your university and government benefactors will be unhappy. Science and Sense isn't nearly as profitable as PseudoScience and Hype.

Okay, bring it, indignant MMCC acolytes. Bring your "peer-reviewed" articles, your "97% of scientists" lie, and your personal attacks. I just put on an extra coat of Teflon. Common Sense and actual observations always trump inscrutable models and "peer-reviewed" baloney.

Climate Part 3

In this third and final column about "climate change", we will explore why the warmists continue with this charade, and list a few of their more laughable predictions. I am certain that my first two columns on this topic have sealed me into the tomb of "not a scientist!" in the warmists' minds, but sorry not sorry, this column is about Science and Sense, not Pseudoscience and Fantasy.

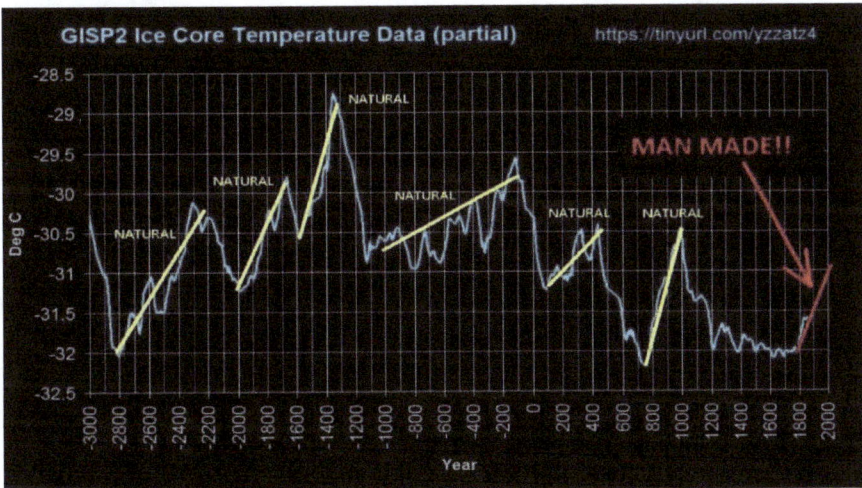

So first, why do they continue pushing this man-made climate change (MMCC) lie, when the rock record (open for anyone to see) shows conclusively that the planet's CLIMATE NATURALLY FLUCTUATES WITHIN A RANGE? Three reasons. 1) The money river from government continues to flow, and writing "peer-reviewed" papers (that can never be proven)

is a gold mine. 2) A business has developed to bilk actual businesses of more money, in the form of "carbon credits", ESG (socialism writ large), and consultants who charge big fees to help businesses bend their knee to the Climate Cabal ("Sustainability" - Snort). 3) Pride. Haughty professors and elected officials are loathe to say "Oops, we were wrong. The CLIMATE IS FLUCTUATING WITHIN A NORMAL RANGE. Sorry for the fear-mongering". Backing down from their always-wrong predictions is not in their ego-driven mindset.

I am not from Missouri, but as an actual scientist, I have always appreciated their state motto, "Show me". Actual science involves real evidence (not manipulated computer models with dubious assumptions). I have traveled all over the world, and I have never seen any of the dire predictions of the warmists come true. I have witnessed hideous localized contamination in India, and the effects of concentrated humanity in London, Houston, New Jersey, Bangalore, LA, and Mexico City. I have also marveled at the pristine beauty of the Canadian Rockies, Hawaii, Scotland, and 1000 places in all 50 states (yes, I have visited all 50). On balance, my personal observations have "shown me" that the Earth is just fine, with a few exceptions in industrial/urbanized places.

But the drumbeat of fear-mongering continues. A 1922 Washington Post article (yes 1922, not 2022) predicted loss of polar ice, rising sea levels, depleted fisheries, and crop failures on a huge scale (NEVER HAPPENED). On the first Earth Day in 1970, the following points were touted: 1) End of civilization in 15-30 years (NEVER HAPPENED), 2) 100-200 million deaths by starvation annually for 10 years (NEVER HAPPENED), 3) A new Ice Age by 2000 (NEVER HAPPENED).

Al Gore, that famous climate expert (snort), predicted in 1999 that Florida would be underwater by 2009 (NEVER HAPPENED). In fact, I first visited Cocoa Beach as a child in 1967, and most recently in 2018. The beach is right where I first found it. Other gems from "Climate Al the Soothsayer" – 1) Big increases in sea levels (NEVER HAPPENED, and he bought a mansion on the beach). 2) A new Ice Age in Europe (NEVER HAPPENED). 3) Massive flooding in India and China (NEVER HAPPENED, just normal flooding from seasonal monsoons). 4) Melting Arctic (NEVER HAPPENED, polar ice coverage actually increasing). 5) Polar bear extinction (NEVER HAPPENED, their numbers are actually increasing). 6) Dramatic worldwide temperature increases (NEVER HAPPENED, fluctuating within a normal range). 7) Hurricane Katrinas every year, with increases in the number and severity of hurricanes (NEVER HAPPENED, all fluctuating within a normal range). 8) Crop failures and mass famine due to shifting climate destroying crops (NEVER HAPPENED, crop yields actually up). 9) A "planetary Climate Emergency" by 2016 (NEVER HAPPENED). I could go on with Al's rantings, but you get the point. The fact that none of them ever happened is an inconvenient truth.

So, the observable Science, and common Sense, indicates that the MMCC lie has been debunked. Yet the warmists continue on their fear-mongering parade, although more and more of us are pointing out that the Emperor has No Clothes (Hans Christian Anderson was wise). Why do they continue with their fairy tale? Money, for one. Power, for another, to push us toward globalism.

What better way to make countries abandon their sovereignty (and their citizens' freedom, and their treasure) than to convince them that they must all "work together" and "make sacrifices" to "save the planet"? A one-world climate agency could govern all, with the stated goal of keeping weather at bay (with the actual goal of controlling free countries, and redistributing their wealth). Not to go too far down this road, but think about it – who stands to benefit from shutting down industry? Government officials and their lackey Climate Academia, whose influence and power would be elevated, with free market businesspeople crushed. Communist countries like China just laugh at the climate hubbub, knowing that globalism just hastens their control over all of us.

So, after three articles, I have covered how "man-made climate change" fear is ginned up, how the government-academic axis validates the lie, how anyone can see that it is all a fiction, and most importantly, why we are being led down this destructive path.

To close out this topic, I urge you to think for yourself. Do your own research. Question the climate change assertions. Follow the money. OR ALTERNATIVELY, go forward knowing that Mother Earth is taking care of the atmosphere, land, and oceans, and Man will adjust to any minor changes in climate. Push back when you can, either by spreading some Science and Sense, or just talk to other people who think, and don't just blindly accept. Voting matters, too – consider supporting people who actually think, rather than blindly follow an unsupportable dogma.

Whenever I am faced with an unhinged, sad, panicky climate nut, I quote Scarlett O'Hara – "Fiddle-dee-dee", and go about my day enjoying our mountains, streams, clean air and water, and abundant wildlife. Life is too short to spend it worrying about a unicorn attack.

The Science of Wasting Brains in College

This follow-up column to the Climate series (which caused lots of discussion, both in favor and against) covers a side topic that warrants attention. No, it isn't about wasting brain CELLS in college (which is a common theme for collegians). This column is about wasting more precious things, like money and time (each worth spending wisely), research dollars, instructional time, and student's careers. In the many articles and sources that I researched for the Climate series, and in the items given to me after the series by interested parties, one particular item sparked this column. A friend gave me an alumni magazine from a prestigious university in Southern California, focusing on the Arts and Sciences graduates, and prepared by (interestingly) the University's marketing department. This particular issue was called "The Energy Issue".

Most of the articles covered student research activities, energy topics, and interestingly, an article about how to gin up man-made climate change "support" by manipulating emotions. All of the articles had several references to "climate change", and "man's role in it". Lots of words were spent reinforcing the underlying dogma of academia, which blames climate changes (natural) on man's activities (untrue), and how the university is spending all kinds of grant and endowment money "studying it". Much "energy" was spent in writing all these articles to reinforce the lie of man-made climate change (MMCC), which justifies how the university manages to blow through hundreds of millions of "research" dollars per year.

As I read this glossy-paper propaganda piece, all I could think was "what a waste!", and not just the glossy paper, which I know is expensive stuff. This university, over the last 100 years, has done wonderful research work to solve REAL problems in our world. They used to have a top-flight Geology program. So I know they have the capability to do meaningful work. But the really tragic waste, not only in dollars, is the wasted brainpower of professors and students who chase this MMCC fairy tale.

I think about these students, getting degrees in silly things like "Sustainability", which just means making good decisions when designing the built environment. Any good engineer or scientist already does that. I think about students with high-functioning brains, getting indoctrinated in this MMCC-based "science", which is based on lies. Like a building built on quicksand, their careers (and self-esteem) will someday crumble when they suddenly realize that they've wasted their money, time, effort, and sweat on a topic that, in the end, their efforts can do nothing to change. And a degree in "Sustainability" has no cross-merit, i.e. it isn't useful in other disciplines. In other words, a degree in "Sustainability" is not sustainable.

I pondered why universities push students in this ultimately useless direction. Of course, they want the tuition, and they need bodies to do the leg work of spending federal grants (on how gas stoves or ceiling fans will destroy the world, snort). Then I had a thought that was like a light bulb going on, but a cold, chilling one that revealed a sad reality. Perhaps universities spend their time, blood, and treasure on the phantom MMCC because universities and industry have already solved the big problems, and now all they have left are imaginary problems with no solution.

Think about the big problems that research has already solved – clean, safe drinking water for the masses. An abundance and dizzying variety of inexpensive, nutritious food. Electricity at our command (well, except for the failing grid, which is another column). Transportation, by multiple modes on land, sea, and air, available to almost everyone. Communication, both mass communication and interpersonal communication, via multiple modes (perhaps too many!) Rapid and effective cleanup of contamination when it occurs. A plethora of drugs for every possible ailment (although cures for several killer diseases still elude us).

Could it be that academia is engaged in the trivial, imaginary pursuit of "man-made climate change" because that is all that is left? Or because it is an easy out – a "science" that can't be proven?

Nearing the end of my 40-year career (5 in petroleum exploration, and 35+ in hazardous waste assessment/cleanup and infrastructure projects), I look back at a body of work that provided energy for human betterment, left the Earth a cleaner place, and that was 85% fun for me to do (I only wish everyone could enjoy their jobs as much as I have). I am concerned about graduates today that have been duped into careers that, at the end of their 40 years, won't matter. I can only imagine the emptiness that those kids (adults by then) will feel.

So, if any college students are reading this, I urge you to consider career paths that are meaningful. Engineering (hey, I built that)! Sciences (hard sciences like geology, chemistry, biology) (hey, I helped build that)! Medicine (hey, I saved you)! Law (hey, we sued the pants off that bad guy)! Or even better, pursue any number of career paths that do not require a $300,000 piece of paper from a four-year indoctrination center. I am forever indebted to plumbers, electricians, carpenters, bricklayers, painters, cooks, servers, mechanics, HVAC techs, fence builders, roofers, etc. for my daily comfort and convenience. And you know what, some of the happiest people I know are in that second group, the ones without degrees.

Bottom line – don't waste your brain, and think about what your 40-year career might look like at the end, as you reflect back on what you did. Spend your brain wisely, you only have one.

Superfund Part 1 – In The Beginning. . .

After winning World War II, the United States saw an explosion in industrial production. We had developed an immense and varied manufacturing capability for the war effort, and most of these facilities were quickly re-purposed to make consumer products. Post-war America had a thirst for the good life, and rapidly growing families and suburbs needed "stuff". Cars, refrigerators, radios, furniture, cribs, building materials, airplanes, tires, you name it. Through the 1950s and 60s, and up until the mid-1970s, industrial companies ran at full tilt making things, keeping people employed, and trying to do so for maximum profit. Waste products were dumped in the river, buried out back, or swept under the rug, since there was no profit in them.

Rachel Carson's 1962 book "Silent Spring" was an early wake-up call that industrial pollution was altering our ecosystems. By the mid-60s, people were noticing things like a dead Cuyahoga River in Ohio (an industrial waterway that actually BURNED in 1968). The Houston Ship Channel was also a chemical cesspool that regularly caught fire. Bald eagle and California condor numbers began to drop, as the pesticide DDT thinned the shells of their eggs. The "hippie" movement grabbed hold of the environment as one of their touchstones, helping to establish the first "Earth Day" in April of 1970. The famous "crying Indian" TV ad helped to turn public opinion, making the average American concerned about how we treated our Earthly home.

Beginning at the end of the 1960s and gathering momentum through the 1970s, Congress acted to pass a raft of environmental regulations. The National Environmental Policy Act (NEPA) established a framework for approving public projects, considering a BALANCE between man's needs and preservation of nature. The Clean Air Act gave us underpowered cars, but cleaner air, and smokestack emissions limits for industrial polluters. The Clean Water Act, Safe Drinking Water Act, and others passed quickly, making Richard Nixon the "Environmental President". Later in the 70s and early 80s came the Resource Conservation and Recovery Act, which created management tools for hazardous materials usage and tracking.

In the late 1970s, something was found to be amiss near Niagara Falls, New York. A cluster of childhood leukemia cases and other rare cancers cropped up in the Love Canal neighborhood. A former industrial waste dump, originally excavated as an unfinished extension of the canal system in the area, was found to be the source of the problem. The hole was filled in the 50s and 60s by several industrial polluters, with dumped drums, or in some cases tankers which released their waste directly into the hole. Later, a neighborhood was built on the edges of the then-filled-in hole, including an elementary school and its playground. This column doesn't have room for

the whole sordid story, so look it up. Or take me out for a beer – I know some of the project managers on the Love Canal cleanup, and their stories will curl your hair.

Dying children and abandoned neighborhoods got the attention of the press, and the American public demanded that Congress act again. The Comprehensive Environmental Response, Compensation, and Liability Act (CERCLA) was passed in 1980. This law allowed for funding (through fees from all companies that manufacture hazardous substances) for cleanup of abandoned hazardous waste sites, among other regulatory tools. The fund generated by the mandated fees was called "Superfund". Overseeing Superfund cleanups was the original and primary charge of the Environmental Protection Agency (EPA), a cabinet-level agency that is independent of other governmental structure (this will be important in the next column).

Initially, the EPA blitzed every state in the early 1980s, identifying, ranking, and cataloguing sites for potential inclusion on the National Priorities List (NPL, Superfund's ranking system). With limited early funding, the EPA ranked sites for their immediate risk to the public, and spent their money and

70

effort on the worst offenders. These included Love Canal, Valley of Drums in West Virginia, the afore-mentioned Cuyahoga River in Cleveland, certain uranium mining sites (Uravan/Naturita, Colorado), and numerous other sites, including Department of Defense sites.

Lower-ranked sites were monitored, and cleanup depended on the site's ranking and available funding. Many sites were "de-listed" from the NPL, with further cleanup passed on to other agencies or state-level programs. Most of the original NPL sites in Arizona were either de-listed or moved to adjudication by the Arizona Department of Environmental Quality (with both success stories and failures).

Now, 40-50 years later, we are reaping the benefits of these early, comprehensive, effective environmental protection laws. Our air is clean (compared to 1970), as is our water. Our cars have recovered their horsepower (and more), as engineers learned to adapt to the new emissions requirements. Bald Eagle and California condor numbers have spectacularly recovered. Industrial polluters found other ways to make their products with less toxic options, even turning some waste products into new products, through the miracles of chemistry. We are a smart and adaptable species.

The picture is not all roses and rainbows, however. Many Superfund sites remain untouched and toxic, either because there isn't enough money in the world to clean them all up, or the technology doesn't exist to do so. In those cases, risks are reduced by managing the site and minimizing human and environmental exposure. Some new technologies are toxic time bombs waiting to happen, like EV batteries, solar panels, and wind turbine blades – all non-recyclable, and destined to fill a thousand landfills in the near future. Miles to go before we sleep.

But there is room for optimism, and pride for our accomplishments in keeping the US clean, with a growing population and an ever-more complex suite of chemicals. The next two columns in this series will cover how Superfund has been politically hijacked, and how we, as citizens, can demand that they keep to their original, noble goals.

Superfund Part 2 – A Hijacked Mission

In the first column of this series, we covered the reasons why the EPA and the Superfund program were needed, how they came into being, and how they were funded. This column will focus on how the EPA and the Superfund program have changed, in many ways not for the better, in recent years.

Once the Superfund program was established in the early 1980s, the EPA went about their task (cleaning up abandoned hazardous waste sites) with due diligence, getting as much done as they could within financial and technical constraints. The definition of "financial constraints" is simple – available money. The definition of "technical constraints" is equally simple – available technology to effect cleanups. But starting around 2008, another constraint came into play – politics.

From the early 80s until about 2009, the EPA operated effectively under Presidents of both parties. The EPA (and it's charge, the Superfund program) stayed laser-focused on their mission – cleaning up abandoned hazardous waste sites. EPA was also tasked occasionally with providing expertise in other initiatives not directly involved with cleanup, because of the scientific and technical expertise within the agency. The EPA should be commended for maintaining high performance on hazardous waste site cleanups, even when their focus was sometimes distracted by other initiatives. The EPA (mostly their private-sector contractors) became effective in executing cleanups. That said, EPA is still a governmental agency, and there was a lot of waste and inefficiency in their processes.

An oddity of the EPA is that it was not created by, nor has direct supervision by Congress. The EPA came into being through an Executive Order by President Nixon, and as such, the EPA answers first to the White House. Way back in the early 70s, when Presidents only occasionally issued an Executive Order, it was done on an issue where there was broad, bipartisan agreement. Cleaning up our fouled nest was a popular topic for both parties in the early 1970s, because it was obvious that action was needed. Funding still goes through Congress, but Congress-critters have gotten very good at hiding money in funding packages for all sorts of pet projects. Sometimes funding for cleanups (or other initiatives) became part of the pork process, where a Congress-critter would sneak cleanup money into some other bill (say, for "education", or for "inflation reduction") (snort). In other words, scientists at EPA deciding where the cleanup money went became less clearly based on severity of the problem, and sometimes the ego of a Congress-critter came into play.

The EPA operated for about 25 years this way, mostly effectively, and mostly focused on their mission. But then along came 2009. In a very short time, the EPA became weaponized, and much of their attention was diverted away from abandoned hazardous waste site cleanups. The false narrative of "man-made climate change" came into the picture, and suddenly the EPA became weather experts (don't we have NOAA, NWS, and about ten other federal agencies with more weather and climate expertise?) You might ask why the EPA suddenly got thrown onto the left's "climate change" bandwagon. The answer is simple. As a Cabinet-level federal agency, answerable to the President and only indirectly by Congress, a President with an agenda can hijack the EPA. And the EPA has regulatory authority, taxing authority, and the ability to use lawyers and fines to make their point. In other words, the EPA was changed from a basically scientific agency to a version of Mafia enforcers for a "climate change" agenda.

And it wasn't just "climate change" that EPA suddenly became interested in. Things like whether you can use the rainwater that falls on your land. Like whether a mostly dry ditch in your pasture can be regulated as a "navigable waterway". Restricting tailpipe emissions to the point that ridiculous EVs become the only form of auto propulsion allowed by a dictatorial government. The EPA became weaponized, so that a rogue executive branch could act like a dictatorship, sticking its federal nose into issues that were none of their business, and punishing select enemies.

Meantime, one side will say that it isn't a zero-sum game, they will just add more people to EPA, adding to the ever-growing cancer that is the overreaching federal octopus. I would direct these people to the Tenth Amendment, and to the Constitution, which limits Federal powers to JUST those enumerated therein. The Tenth Amendment clarifies that states are responsible for everything else. If we are going to take that seriously, we should disband the EPA, take maybe half their funding allocation, and divvy it up among the states by population, saving the rest and lowering the taxpayers' bill. It has been this author's experience that state-led cleanups are far more effective, efficient, and timely than federal Superfund cleanups, anyway.

The EPA, like unions, started as a good idea. But they slowly became politicized, and no longer effectively serve the function intended. Politicians are very good at setting up new government "solutions" to problems, and very bad at dismantling solutions, groups, or agencies that have lost their way.

If the EPA were to step back from meaningless political initiatives (like "man made globull warming"), and focus on what they were good at for many years, contaminated sites that have been sitting dormant for decades might finally get cleaned up. The EPA, if refocused on their original mission, could likely function on half their current budget, or less. And in today's tattered economy, it would be great to see even one slice of government do what Americans have to do every day – watch their budget, and spend only where it counts.

But I'm not holding my breath.

Superfund Part 3 – Our Local Superfund Site

The first two columns in this series covered how the Superfund program for cleaning up abandoned hazardous waste sites came into being, and how the US EPA has become distracted and weaponized by an out-of-control federal octopus. This column brings the Superfund program home, to our very own site right here in Yavapai County – the Iron King Mine / Humboldt Smelter site (IK/HSS) in Dewey/Humboldt! The site itself consists of 153 acres including the old Iron King mine, a site at the east end of Main Street, and 182 acres including the old Humboldt Smelter and part of Chapparal Gulch (you can see a map of the site on EPA's website).

The Iron King Mine and Humboldt smelter are typical of mining/smelting operations of their time. The smelter came first (with crushers operating even before the smelter), serving ore production from other area mines from the late 1800s through 1937. The smelter took crushed ore, heated it in a coal-fired crucible, and produced metal ingots that were sent away by rail for further processing. Gold, silver, zinc, copper and lead were produced at the smelter. The Iron King mine was located nearby, and operated from the early 1900s until about 1970. The mine operations resulted in about 4 million cubic yards of orange-colored waste tailings (orange from the rust [oxidation] of iron, entrained in the matrix), which were piled near the mine and along Chapparal Gulch. The smelter generated slag and dross (the crust that is left over after the crucible-firing to extract the metals). The piles of slag/dross were located near, and mixed in with, the tailings from the mine.

Contamination did not stay put in the tailings and slag/dross piles, however. The smelter's stack distributed heavy metals across a wide area, including primarily lead dust. Rain events washed wastes into Chaparral Gulch, and eventually downstream to the Agua Fria River. Wind carried dust from the piles across a large area, including the Dewey/Humboldt town site. A dam was later constructed along the lower Chaparral Gulch to limit movement of the waste products toward the Agua Fria.

Mine sites and waste pile areas will always have elevated metals in soil, that is just a fact of mining. The town of Dewey/Humboldt is impacted in this way, like hundreds of other mining towns in the US and elsewhere on the planet. Some of these metals are toxic in elevated concentrations.

EPA's approach to the IK/HSS has been typical for sites of this type. The goal in cleanup is not to remove ALL contaminated material – there isn't enough money in the world to completely clean up thousands of mine sites. But since this Superfund site is located where people have chosen to live, the EPA has engaged in a risk-reduction approach.

Initial remedial efforts in 2010-12 and 2017 focused on removing impacted surface soils in residential areas, and stabilizing source areas for blowing dust. Recently, a plan for site remediation was finalized. This plan will continue to stabilize dust sources, augment runoff control in washes, and relocate materials to two controlled, landfill-like cells. These cells will be located near their source areas.

These landfill-like disposal cells will concentrate the currently open-dumped tailings into lined, covered, monitored cells, very much like a municipal landfill. Leach water that will flow through the cells will be collected and treated, capturing the contaminants prior to release of the treated water. This kind of "encapsulation and source management" approach has been used at hundreds of mine sites in the US, and is effective at reducing the risks to humans and the environment from these wastes.

Of course, like any cleanup plan, some people like it and some people don't. For the IK/HSS, some locals want more cleanup, and for the waste to be shipped further away. This is understandable, but remember, they knowingly bought homes near an old mine, what did they expect? And total cleanup is not cost-feasible, even if it isn't their money (It is OURS, since federal Superfund cleanups are done with a mix of tax dollars and industrial fees). And moving the waste cells further away introduces more risk and cost – trucking these wastes risks contaminating more areas with the waste, fugitive dust, or releases from a truck accident. Generally, the EPA chooses to place waste cells as close as possible to the source, preferably on areas within the site that are suitable for such a facility.

Put simply, the approach at IK/HSS will put the mess into a big plastic bag, bury it, and monitor water that percolates through it, treating it before release. Just like most mine sites, the goal is human health exposure risk REDUCTION, not elimination. The plan aims to reduce risks and exposures to levels below human health impacts. The EPA team employs toxicologists and other environmental exposure experts, and this author can attest that the plans that EPA implements, after years of study, are designed to succeed, not fail.

I have visited Dewey/Humboldt, and will continue to do so. It exists for the same reasons Bisbee, Jerome, and Ajo AZ (plus Silverton, Leadville, Creede, and Central City, CO, to name just a few) and hundreds of other old mining towns exist. They are loaded with charm, often in a beautiful setting, with interesting history and residents who tend to be characters. Most of those residents know that the mine existed there, and that the orange dirt is part of the mining process. They choose their risks, just like Oklahomans choose to live in Tornado Alley, and Californians choose to live where the ground shakes. At least in mining towns, you can take steps, as can the EPA, to protect yourself. In Moore, OK or Oakland, CA, the EPA can't help you with your risks.

Common Sense Environmentalism

If you are a regular reader of this column, you are likely aware that Common Sense and certain current fads (like EVs, "man-made Climate Change", and silly roundabouts) don't really intersect. Any issue, old or new, should be evaluated through the tried-and-true lens of common sense. If something sounds odd, off-the-wall, or made-up, it likely is. Observational science is powerful, and is made of common sense. For example, I don't need a computer model to tell me it is hot in the summer and cold in the winter, and that some years are dry and other years have more precipitation. So, if you want to be a good steward of the planet, apply common sense in your daily lives. Doing that makes it possible to still be an "environmentalist", without losing your mind.

One great example is water. We live in a desert, so even though water is always at the ready in our tap, we should know where that water comes from, and treat it with respect. We all want our creeks and rivers (such as they are in a desert) to remain as wet as nature allows. The Upper Verde River, a source of much fear-mongering about it running dry, remained wet through our 26-year drought, even as population has grown during the drought. The seven Salt River Project (SRP) lakes, critical to our state's supply, stayed wet even in dry years (through good stewardship of the resource). Living in the desert, even in an urbanized setting which provides for our water needs, requires respect for our limited wet resource.

There are things that you, as an individual, can do that really help reduce our water footprint - small things, that just take getting in the habit of doing. Turn off the water while you brush your teeth or shave. Try taking shorter showers (even 10% shorter helps). Run the dishwasher rather than hand-wash (dishwashers use 40% of the water of hand-washing). Run only full loads in the washing machine. Capture rainwater off your roof downspouts in a barrel, and use the rainwater to water the flowers and gardens. Replace thirsty grass lawns with xeriscape (xeriscaped yards can also be lush and green, if you choose the right drought-resistant plants). Use an automated car wash – they reclaim/reuse most of their water. If we, as residents of the Greater Yavaplex, can use 10-15% less water, that delays having to expand our groundwater draw by years (and saves you money!)

Heating and cooling is another area where you can make small changes that help the planet (and save you money!) On hot days, use a fan in the room you are inhabiting. Moving air feels much cooler on the skin, and you might be able to delay turning on the A/C, or maybe not turn it on at all. In the winter, catch the passive heat from the sun by opening blinds on the east side of the house in the morning, west side in the afternoon. Then at night, be sure to close all the blinds (heat escapes through glass). Replace that old, worn-out weather stripping on doors and windows to keep that heat in. And make sure your filters are clean – an AC or heating unit has to work a lot harder to push air through a clogged filter.

Food is another item that is resource-intensive. For example, consider the tomato. If you buy a nice beefsteak tomato at the store ($2 to $3 each, yikes), think about the resources it took to get it to you. Water at the commercial farm. Trucks and labor to bring it in from the field (fuel). Trucks to get it to your store (more fuel). Labor to stock it on the shelf. Rent/taxes/utilities for the grocery store space. Labor to check it out (okay, likely your own labor, since grocery stores have turned us all into checkout clerks). Now, contrast the expenditure of all those resources with the garden-grown tomato in your back yard. It still requires water, but all those transportation and labor steps go away. You just pick it off the vine and eat it. And it tastes better, too. In our generally poor soil in Prescott, preparing a garden takes some up-front work (we use raised beds). And you have to commit to watering, weeding, covering when it's cold, etc. The missus and I really enjoy time in the garden – no media or politics! A garden will save you money, and you will tread lighter on the Earth. And you can garden year-round with indoor lights, or a hydroponics setup!

These are just a few of the little things you can do to be a good steward of the planet. Writ large, these steps are referred to as "conservation". The root of that word is "to conserve", or to protect something of value. Being conservative MEANS taking good care of something you care about. The last thing you want to be is liberal – liberal with your use of water, liberal with energy resources, liberal with fuel burned to bring you food. The intangible benefit to being conservative is the great feeling you get, knowing that you are treading lightly on the planet, and keeping it in its best condition for kids, grandkids, and generations 300 years from now. And you don't even have to wear tie-dyed T-shirts and Birkenstocks (unless you just want to)!

CHAPTER 3
Energy Topics

INTRODUCTION

Energy is a broad topic, and I have professional experience with almost all of the energy sources that are covered in the five-part energy series. A huge amount of misinformation is out there about all five of the sources covered, and the goal of the series was to lay out a brief description of each, with pros and cons for each kind. The reality is that any form of energy, harnessed for man's use, is a tradeoff between the benefit gained by liberating and using the energy, and the damage to the planet caused by harnessing it.

Following the five-part Energy series is a two-parter on a side topic for petroleum, hydraulic fracturing. Again, tons of misinformation exist about "fracking" (a stupid-sounding contraction of the term). Once you understand the mechanical engineering and the geology of hydraulic fracturing, any thinking person would understand that it is a harmless way to recover WAY more petroleum from any well.

A column about the current fad of Electric Vehicles is included, and I have to say that in 50 years, if somebody finds this book, they will probably laugh at the inclusion of this topic, since EVs will be a long dead and forgotten boondoggle by then.

In a related topic, this chapter finishes with a column about our overstressed and aging power grid – something that is as serious as the EV fad is hilarious.

I suggest that the reader "fuel up" on a couple of cups of strong coffee before diving into the "energy" topic.

Energy series #1 of 5 – Coal

A "hot" topic these days is how we fuel our modern society's needs. The next five columns will be a weekly series about the energy sources that we smart primates have developed for our comfort and advancement. These columns will discuss each energy source's "pros and cons", hopefully to prepare you better for any friendly discussions. These columns also will present energy sources in a sort of chronological way – from **coal** (better than a wood fire), to **hydropower** (natural and renewable), to **petroleum** (the powerhouse of the modern world), to **nuclear** (controversial but extremely efficient), to **alternative** (wind and solar). As preamble to this series, we North Americans should be ever thankful that we live in North America, easily the richest and most diverse continent for energy sources. We are endowed with giant supplies of each type of energy. This diversity of energy wealth allows us even to have this discussion at all. So first, let's talk about **coal**.

First off, what is coal? It is a hydrocarbon fuel, meaning that it contains the same basic chemical stuff as petroleum, but in a soft-rock form, rather than a liquid. Plants and critters die, their bodies get buried in mud, and enormous amounts of time pass. The hydrocarbons decompose, get crushed by enormous geologic pressures, and recombine into the black chunks that we mine as coal. Several grades of coal exist, from the most energy-dense (anthracite, think full-sugar Coke), through bituminous (watered-down Coke), to lignite (diet Coke), to peat (barely a fuel, popular in India, more often a fertilizer). Coal generally occurs in horizontal "seams," sometimes close enough to the surface to "strip mine" (a process that removes the soil above, leaving a scar on the land). Deeper deposits require more traditional mines (with shafts) to bring the coal up.

A big advantage of coal is that it is well-distributed in North America, close enough to be easily transported by train to every point-of-use. A big cost component of any energy source is transportability to its point-of-use (this will be part of every discussion in the Energy series). Back when coal was a household fuel, trucks (and wagons before that) would bring the coal to individual homes and businesses for end-point use (furnaces, boilers, etc). After consumers switched to more convenient energy sources (like electricity and gas), coal was used as a more centralized energy source, fueling coal-fired power plants, producing electricity. This is the main use of coal today. Coal's energy use in Arizona stands at 13% of the mix, in the US at 16%, and dropping. The US is third in coal power usage, with India about 20% greater than the US, and China *6 times* our usage (and growing rapidly).

PROS – Coal is plentiful. Even though we have already mined most of the energy-rich anthracite coal, mid-grade (bituminous) coals are still abundant. We have an existing (*paid for*) rail system and power plants to move and extract the energy from this resource. We have a mining workforce (concentrated in certain parts of the country) who not only know how to extract coal, but are itching to get back to work (see **CONS** below). We have existing mine sites that can supply our needs for decades, without disturbance of pristine areas. And we have centuries more coal available if we choose to go into those pristine areas. Coal is also cheap, with a *very* low cost per unit of energy.

CONS – Coal is "dirty." This means that burning coal as a fuel liberates a lot of nasty byproducts that aren't healthy for living, breathing things. Sulfur is emitted from stacks, going straight into the atmosphere, and recombining with water vapor to make sulfuric acid (acid rain). Many other carcinogenic chemicals come out of those coal-plant smokestacks, and we have yet to devise an effective way to remove all the bad stuff from coal plant emissions. Smokestacks are required to have "scrubbers" which remove most particulates, but we have a long way to go before coal can even be considered close to "clean." (BTW, there is no such thing as "clean coal", a political lie that was a red herring for the industry. The regulations passed for "clean coal" emissions goals were unattainable, effectively curtailing coal use). Coal mining is also destructive at the mine

89

sites. Reclamation of mine sites is required at closure, but really, the mined area will never return to what it was before mining. The final big negative with coal is the health effects on living things. The miners themselves are most affected through direct contact, but the public in general (and animals) suffers detrimental effects from acid rain, degradation of air quality, and pollution of water resources. And it gets worse the closer you live to a coal-fired power plant.

SCORECARD FOR COAL – From a financial standpoint, coal is a positive. It is cheap, we have copious amounts of it, we have infrastructure already built to use it, and we have people that need the jobs to extract it. But from an environmental standpoint, coal is a pretty bad choice. All the technology we can bring to bear will not eliminate the nasties that are a byproduct of coal use. But we can certainly make strides to clean up what we can, and manage the wastes responsibly. Now if only China and India would do a tenth of what we do in responsible coal use . . .

So, my *bottom line* on coal is thumbs-sideways. We will certainly need coal as part of our energy mix going forward (for decades), but we have a lot of work to do to minimize the environmental damage from using it. Coal should eventually be phased out, but we aren't anywhere near that point yet.

Next week – Hydropower (dam dams)

Energy series #2 of 5 – Hydropower

This is the second column about various ways we fuel our modern society's needs. This series covers the five major energy sources, including "pro" and 'con" arguments for each, to better prepare you for friendly discussions. First we covered **coal**, now **hydropower**, next **petroleum**, then **nuclear**, and finishing with **alternative fuels**. As a preamble, we North Americans should be ever thankful that we have rich reserves of each of these energy categories. This diversity of energy wealth allows us to even have this discussion. So without further fanfare, let's talk about **hydropower**.

First off, what is **hydropower**? It is the generation of power (electricity) from harnessing the kinetic energy from falling water. In order to make hydroelectric power, you need to start with two basic ingredients – flowing water, and elevation differences. You also have to be able to stop the water, and funnel it through a device to catch the force of the falling water. Dams are a dandy way to do this, although early hydropower applications included waterwheels, which caught just enough kinetic energy to power maybe one farm, or a mill. Modern hydropower involves large dams, with water/energy storage behind them (technically reservoirs, but I will use the common vernacular "lakes"), from which water is funneled through the dam, then through large turbines. The falling water turns the turbines, which have copper coils and magnets in them that generate electricity when whirled past each other. The electricity is then sent through large power lines to distribution lines, and eventually into your home or business.

Hydropower has big advantages, and a few big problems. First, you need a dam, on a river, in a canyon. A large river and deep canyon provide the best opportunity for energy storage and generation. Hoover Dam is a great example, which impounds the Colorado River. Big, big river with a deep canyon. Hoover Dam is such a great setup that it has 17 penstocks/turbine sets, with a maximum output of over 1300 megawatts of power. Right now, power production is way down, though, since Lake Mead is over 160 feet low (~30% capacity) due to a prolonged drought. But still, Hoover Dam provides reliable, non-polluting power to the southwestern power grid. Power that is relatively cheap – costs for construction of the dam have long since been recovered.

Some drawbacks exist though, in that a canyon was drowned and lost forever (Boulder Canyon), along with several towns which lay below the elevation of the finished lake. This is true for all hydropower lakes – they drowned a canyon in favor of a lake – and sometimes a beautiful canyon. Hydropower, like all energy sources, is an exercise in tradeoffs. Hydropower accounts for 5% of our power in Arizona, and about the same nationally.

PROS – Non-polluting, long term (over 100 years for most dams), reliable power, at a very low cost. Recreational opportunities on the new reservoir, although some would argue that recreational opportunities on the river were lost – maybe a tie on that argument. Flood control – in the Hoover Dam example, the dam and lake stopped centuries of devastating floods along the lower Colorado River. Irrigation for farms, possibly in areas where farming was impossible before, due to low rainfall (again, the Hoover Dam / Lake Mead example). And reservoirs provide reliable water for human use.

CONS – The energy is not diversified – you certainly can't power your Southwest Airlines flight's engines with a dam. Dams are limited to electrical power production. A canyon is lost, forever. Some environmentalists argue for removal of the dam and restoration of the canyon, but decades of silting would have destroyed the canyon, and it would take centuries, if ever, for the valley to recover to anything like it was before. Loss of town sites along the river, below the lake elevation. Maintenance costs, huge when the turbines need to be replaced (about every 100-125 years). Limited reach and location – clearly Portland, Oregon and Las Vegas benefit directly from hydropower, but it is a hard sell in Wichita. Geography rules the hydropower industry. And finally, negative impacts to fish in some rivers. The Pacific Northwest is what I call "fish crazy," in that every public expenditure decision they make has to include potential impacts to fish populations. Dams aren't good for fish migration patterns (duh), although the engineered "fish ladders" at Bonneville Dam are a must-see if you are ever up there. Fish issues can be mitigated, but to some, fish impacts are a detriment to hydropower. Also, with the rise of modern, lawyered-up environmental activist groups, it is a safe bet that we have seen the last major dam construction in our country. So, a con here would be lack of expandability for hydropower.

SCORECARD FOR HYDROELECTRIC POWER – The scorecard on hydropower is a qualified thumbs-up. If the geography works in your area, the dam and infrastructure already exists, and the fish lawyers haven't shut you down, good for you. Hydropower is a boon to your area. The downers are that it is geographically limited, limited to just electrical power generation, not expandable in the current litigious environment, has caused the loss of a likely beautiful canyon, and your local dam probably needs to replace turbines soon (given the age of most hydropower dams in this country).

Hydropower is certainly part of our energy mix, though limited in flexibility and geography. It is certainly clean, cheap to operate overall, and the lake is probably all anyone alive now remembers (bye bye canyon). If we consider hydropower in the larger buffet of power sources, I like it a lot, and wish there was some way to have our cake (the dam, lake, and cheap power) and eat it, too (the beautiful lost canyon). But I know I will certainly lose any argument with a fish lawyer.

Energy series #3 of 5 – Petroleum

This is the third column about various ways we fuel our modern society's needs. This series covers the five major energy sources, including "pro" and "con" arguments for each, to better prepare you for friendly discussions. First, we covered **coal** and

hydropower, now **petroleum**, next week **nuclear**, and finishing with **alternative fuels**. As a preamble, we North Americans should be ever thankful that we have rich reserves of each of these energy categories. This diversity of energy wealth allows us to even have this discussion. So without further fanfare, let's talk about **petroleum**.

First off, **what is petroleum**? It is a hydrocarbon compound, meaning that it contains hydrogen and carbon atoms, arranged in different ways to create a myriad of useful chemicals. Some of these provide us with energy, while others provide the building blocks of plastics, asphalt, pharmaceuticals, and a million other staples of modern life. It comes from plants and critters that die, their bodies get buried in mud, and enormous amounts of time pass. The hydrocarbons decompose, get crushed by enormous geologic pressures, and recombine into a liquid form (oil) or gaseous form (natural gas). Petroleum occurs at all different depths, from near the surface (like the La Brea tar pits in LA), to miles deep (some productive zones are over 5 miles below the surface). Drilling wells is the most common way to recover the oil, and modern techniques such as horizontal drilling and hydraulic fracturing have opened up vast "new" sources in shale rock that we couldn't recover before. Petroleum is also a feedstock for a huge number of chemicals, serving the plastics, pharmaceutical, and industrial markets. But this series of columns is focused on energy sources, so the focus here is on petroleum as a fuel.

Petroleum has a three-pronged case as the supreme energy source for our planet – 1) it is abundant (but not distributed evenly – we have lots, Japan has none), 2) it is incredibly energy-dense, and 3) it is portable and versatile. Oil is located on every continent, and is easily accessible in some places (like west Texas, the Middle East, and Asia). Other locales have a lot of oil, but it is harder to access (Amazon basin, Antarctica, or unstable countries). Oil's energy density is a huge advantage. It amazes me that one gallon of gasoline (think about a milk jug) can carry my 4500-pound vehicle containing me, three friends, and our golf bags a distance of 23 **miles**. Or, a few thousand gallons of a similarly-refined petroleum can carry 150 people, all their luggage, tons of mail and freight, a few dogs, some drinks and peanuts, and a Boeing 737 from Dallas to LA. Try that with a solar panel, or a bin of coal. Petroleum is also refined into heating oil, kerosene/jet fuel, diesel, solvents, lubricating oils, or a thousand other useful products. Natural gas (also petroleum, just in gaseous form) was burned off as a **waste product** (!!!) for most of my life. Natural gas burns at a very high energy output, with minimal emission of nasty stuff to the atmosphere. North America is also the Saudi Arabia of natural gas - we have more than any other continent. Petroleum and natural gas are also popular fuels for power plants, creating electrical energy cheaply, while generating a fraction of the emissions that coal does. Finally, the US has so much petroleum that just a couple of years ago, we were in a surplus situation, which helps our economy greatly. Petroleum supports ~45% of our electrical generation needs in Arizona (more nationally), and fuels almost all our cars, trucks, trains, and airplanes.

PROS – Oil is plentiful – we have enough to last at least 800 years (maybe forever, since recovery technologies continue to advance). Oil is versatile – many different forms of this energy-dense stuff is used for propulsion, power generation, and heating. Oil is cheap (wait, don't get excited yet) - considering what we get from a barrel of oil, the cost is quite low per unit of energy. We have existing extraction, refining, and distribution infrastructure – much of which is already paid for and in place. We have the most highly trained workforce in the world for petroleum extraction and use. And most of all, our entire economy, including the highest standard of living in the world, is made possible by cheap, locally available petroleum.

CONS – Petroleum, like any other hydrocarbon fuel, creates some nasty byproducts when used. Even our very clean cars of today emit some bad stuff to the air. Although cleaner petroleum fuel options are available for our cars (natural gas fired engines), these cars remain just a fraction of our fleet – mainly because gas and diesel are so well entrenched for cars and trucks. Areas with a high concentration of refineries (Houston, LA, New Jersey) are plagued with air and water quality problems, and have higher-than-average disease rates. Oil spills (from wells, pipelines, trains, ships, etc.) continue to be a problem, albeit one that is happening less often, and we are getting better at cleaning up (trust me, this has been my career for the last 40 years).

So, the **bottom line** on petroleum is an enthusiastic thumbs-up. Petroleum will continue to be our primary energy source for centuries, and we constantly get better at recovering more, and also reducing the contaminant footprint. By the way, clean natural gas is the Next Big Thing (call your broker). I expect lots of outrage from hypocritical people typing on their cellphone or plastic computer, after they get back from driving their gas car to the store to buy lots of plastic items (all impossible without petroleum), while wearing clothing made from petroleum. Fire away naysayers, petroleum wins.

Energy series #4 of 5 – Nuclear

This is the fourth column about various ways we fuel our modern society's needs. This series covers the five major energy sources, including "pros and cons" for each, to better prepare you for friendly discussions. First we covered *coal* and *hydropower*, then *petroleum*, now *nuclear*, and finishing with *alternative fuels*. As a preamble, we North Americans should be ever thankful that we have rich reserves of each of these energy categories. This diversity of energy wealth allows us to even have this discussion.

Since I made so many friends with my Petroleum column last week, let's delve into an even more polarizing energy source – *nuclear*.

What is nuclear power? First, please say it right – NEW-clee-ur, not NU-ku-lar, as Bush 43 liked to say. It sounds dumb that way. Nuclear power is the generation of power (electricity) from chemical reactions in certain radioactive metals in a controlled, shielded reactor. The heat from this reaction is used to heat up water into steam, and turn steam turbines to create electricity. Nuclear power is different from the other types of power in this series, in that it is relatively new (about 70 years old), highly controversial, and has experienced some mishaps that have tainted its use forever, in many people's minds. Many people equate nuclear power with nuclear weapons, which is a complete mischaracterization. Although both power and weapons use atomic chemistry as their basis, nuclear power generation is beneficial, while nuclear weapons, well, aren't (if you are near one during its intended use). Nuclear power is clean (in that it doesn't pollute the atmosphere), reliable (works day or night, wind or no, with no pipelines that can be disrupted), and generates very little waste, compared to its energy output. The downside is that the small amount of waste that nuclear power does generate is incredibly toxic and dangerous.

The world currently has 438 nuclear power plants, with 57 under construction (late 2022 numbers). The US has 81 plants, in 30 states, and Arizona has one of those (the largest of them all, the Palo Verde plant located west of Phoenix). Three major nuclear plant disasters have occurred, which have affected the acceptance of nuclear power greatly. In 1979, a coolant system failure at the Three Mile Island plant near Harrisburg, PA caused a moderate event. Because of that event, no new US nuclear plant sites have been given a permit by the Nuclear Regulatory Commission since 1979 (Arizona's Palo Verde plant is permitted for five reactors, but has only built three). In 1986, the 4-reactor Chernobyl plant in Pripyat, Ukraine literally exploded, causing the worst nuclear power disaster to date. In 2011, an earthquake and resulting tsunami destroyed a

6-reactor plant near Fukushima, Japan, causing worldwide panic (since the site is located on a Pacific beach). These three events permanently colored the future of nuclear power. Unfortunately, we don't hear about the millions of safe operating hours at all the other plants, where things work correctly. All three events were the result of extraordinary circumstances, and all were preventable. We learn from these disasters, and reactor operators consider all of these lessons learned.

A major issue in the US is that we don't have any place to entomb the nasty final waste products from nuclear plants. Well, we do, actually – the Yucca Mountain Facility in Nevada was completed for just that purpose, at a cost of many billions of (your) dollars. But Senator Harry Reid (D-NV), put the opening of Yucca Mountain on indefinite hold due to the political stance of his party. To this day, high level nuclear wastes are temporarily stored at all of the plants where the waste is produced. Some of these sites are located where tornadoes, floods, and other disasters might occur, that are far riskier than the geologically stable and purpose-built Yucca Mountain site. So in the meantime, we watch stockpiles of nuclear waste grow at dozens of facilities, while Washington figures it out. Nuclear energy accounts for 29% of our power generation in Arizona.

PROS – Non-polluting (from an air standpoint), long term (at least 100 years for most reactors), reliable power, at a low operating cost. The US has a lot of our own uranium (fuel) deposits, as does neighbor Canada. Nuclear plants have a small footprint for the site, with the capability to generate enormous amounts of electrical energy. A great alternative for when we run out of coal and oil (which is decades to centuries away).

CONS – Politically difficult to build new facilities, not flexible (only for generating electricity), limited possible locations due to water availability for cooling, very expensive to build, and to decommission at the end of a facility's use, and right now we are politically constipated for managing/disposing of waste material. Many environmental activist groups are dedicated to fighting nuclear power in our country, and they are fully lawyered up, brutal in their tactics, and well-funded.

SCORECARD FOR NUCLEAR POWER – The scorecard on nuclear power depends on your political persuasion. From a scientific standpoint, it is the solution of the future. From a political and public relations standpoint, it is a non-starter (right now, in the US). Thankfully, our existing nuclear power plants contribute significantly to our crazy quilt of power sources. But we better maintain, carefully operate, and protect the ones we have, because we aren't getting more, any time soon.

So, the **bottom line** on nuclear power is a "thumbs sideways now, thumbs up later." It is certainly part of our energy mix, though limited in flexibility and with significant political constraints. It is certainly clean, cheap to operate (until the site has to be decommissioned), and provides many high paying jobs in the communities where reactors are located. But make no mistake, more nuclear power plants *are* in our future. An intangible benefit is the entertainment value in watching a "no-nukes" paid protester go "mano a mano" with a local nuclear plant worker. The red-faced screaming and spitting is epic.

Energy series #5 of 5 – Alternative Energy

This is the fifth of five columns about various ways we fuel our modern society's needs. This series covers the five major energy sources, including "pros and cons" for each, to better prepare you for friendly discussions. We have covered *coal*, *hydropower*, *petroleum*, and *nuclear,* and now we finish with *alternative energy*. As a preamble, we North Americans should be ever thankful that we have rich reserves of each of these energy categories. This diversity of energy wealth allows us to even have this discussion. So, let's put on our Birkenstocks and tie-dyed shirts and talk about *alternative energy* (wind and solar).

First off, **what is alternative energy**? It is any non-traditional energy source. For this discussion, let's limit it to two – wind and solar. Biofuels are out there, but aren't yet viable, so we will exclude biofuels for now. I'm not saying that wind and sunshine haven't been around – they have been since long before we walked this big blue marble. But, relatively recently, modern technologies have allowed us to harness these sources for electrical power generation.

WIND – Actually, wind power has been harnessed for centuries, from Dutch windmills to the American prairie staple we have seen in a thousand watercolor paintings. But now, we have huge wind farms, sprouting three-bladed monsters as tall as a 50-story building. They have big boxes (the size of a travel trailer) atop a giant tapered tower, with three enormous blades to catch the wind. The box holds a mini-generator, which, like the turbine in a dam, spins magnets past copper coils to make electricity. The power then follows collection lines to a transformer, and off to the grid the juice flows. Upside: Once they are up, the source of the power is free (wind). Downsides are that they are energy intensive to build and ship (usually from overseas), and they have a design life of 20-30 years (if everything is well maintained). They also require a lot of maintenance and (petroleum) lubricants. Another downside is that the wind doesn't always blow, so the source is unreliable. And, they tend to chop up birds in flight (now THAT is an inconvenient truth).

SOLAR – The dominant form of solar is Photovoltaic (PV) Cell technology, which directly converts the sun's rays (photons) into electrical energy, and sends it up the line as electrical power. Advantages to this system are simplicity (less maintenance required), and upgrade-ability (when a panel wears out, or a new PV technology comes along, you can just replace the panels). Also, solar panels can be placed in a variety of locations – farm fields, rooftops, mountainsides. The downsides are that both kinds of solar take up a lot of space, and land isn't cheap. Solar installations also strive to be close to existing large-capacity electrical lines, because the tap line to carry energy from a remote location to the grid can be expensive to install. And of course, the biggest drawback to solar is that the sun doesn't shine all the time (like at night, or on a cloudy day, or on a short winter's day). Solar makes the most sense as an off-grid power source for individual homes (coupled with a battery system). Alternative energy accounts for just under 10% of power generated in Arizona.

This is a good place to talk about cost performance. Both wind and solar generate power at very low rates, and intermittently (not consistently). This makes the economics of wind and solar marginal at best. The federal government has provided cost support (*your* tax money) as an incentive to get these technologies off the ground, but eventually those subsidies will expire, or will be rescinded by a more sensible administration, and alternative energies will have to compete on their own merits. Some companies (like Apple, Microsoft, etc) invest in green energy for their plants as a halo thing ("Aren't we wonderful!"). But if I was a shareholder, I wouldn't be excited about elevated energy costs reducing my return, just to feel "green."

PROS – Alternative energy seeks to use "free" sources of energy (sun and wind). They are virtually emissions free (unless you count all the engines needed to build the sites, manufacture the materials, and maintain the facilities). North America is as rich in sunshine and wind as anywhere else on the planet. These alternative energy industries also stretch our minds and ingenuity, always a good thing.

CONS – The two biggest drawbacks to alternative energy are its unreliability, and low energy production rate. The kind of power made by wind and solar is purely electrical power, so it isn't very versatile. Cost to build these facilities, without taxpayer support, is expensive, with an extremely long payback. And of course, windmills chop up birds in flight, spook cattle, and sometimes get hit by lightning and burn. Solar panels take up a lot of space, disorient migrating birds, and will someday fill up a thousand landfills with used/expired (plastic) panels (oh, and their toxic metals).

SCORECARD FOR ALTERNATIVE ENERGY – The scorecard on the various forms of alternative energy is mixed. Taking the politics out of it, the science is very cool, but the economics are sketchy. Particularly wind, a money-loser without our tax subsidies. Alternative sources have an uphill battle against entrenched, cheap, plentiful, and well-marketed energy sources like coal and oil. But variety is a healthy thing, and the crunchy-granola crowd just loves alternatives (despite the fact that they still use oil in the form of plastics, lubricants, and the engines of maintenance trucks). Watch the PragerU video "What's Wrong with Wind and Solar" for an eye-opener.

So, the **bottom line on alternative energy** is a weak "thumbs up" for solar (individual off-grid home use is best), and a thumbs down for wind farms (they just don't work financially without my tax money, and I like birds). Alternative energy will never be sufficient to replace the other sources for our ever-increasing needs.

The Miracle of "Fracking" and
$2 a Gallon Gas (Part 1 of 2)

The United States Geological Survey (USGS) recently announced a petroleum "find" in the west Texas Permian Basin of enormous size and importance (west Texas produces about 40% of our nation's oil). The Wolfcamp Formation (I won't bore you with the specific geology) is a shale formation that has been well known for decades as a huge oil and gas repository. But, as a shale formation, the rocks are dense and platy, so that oil can only flow through them horizontally (and slowly). The USGS estimates that greater than 20 BILLION barrels of oil is found in this ONE formation, making it triple the size of the vaunted Bakken Shale of North Dakota. What makes the USGS announcement pop, is that the oil in this well-known formation, long written off as "un-producible," can now be brought to the surface, through the miracle of "fracking."

Yes, "fracking," the rather stupid-sounding Hollywood contraction of the term "hydraulic fracturing," will bring us cheap oil for generations. But <NEWS FLASH> hydraulic fracturing isn't new. As a young pup I worked in the oil fields of west Texas, including hydraulic fracturing projects in the late 70s, and the technology had been in use since the 1960s. So, let's first establish what hydraulic fracturing is. The practice involves

111

pumping a mixture of water and sand, and a small amount of lubricant chemicals, down the well at enormous pressure. The fluid enters the horizontal cracks in the shale, and basically jackhammers the rock open at a very small scale. This allows oil to flow through the fractured

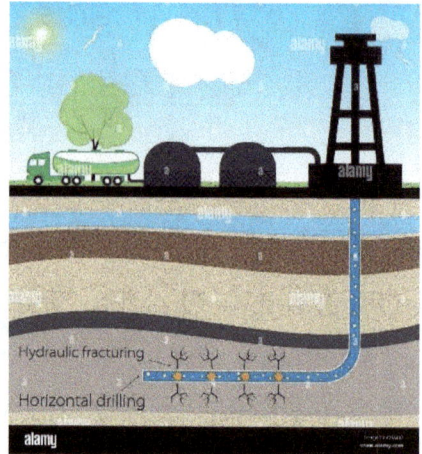

rock, into the well bore, up to the surface, through pipelines to the refinery, making thousands of useful products (including fuel).

So, why is this new find a big deal, if we have had this technology for decades? It is the combination of hydraulic fracturing with horizontal drilling that makes the miracle. Before, the fracturing could only penetrate a short distance from a vertical well shaft. Some oil would be liberated, but after that was pumped, the same slow flow rate problem existed. But then came the game-changer, horizontal drilling. In the past 20 years or so, we have perfected drilling techniques that allow boreholes to be drilled horizontally, ALONG the fracture planes of shale. That translates into an ability to produce that previously "unproducible" Wolfcampian oil and gas (and many other shale formations), in a field where the pipelines, the skilled people, and the rest of the necessary infrastructure ALREADY EXIST.

I helped my Dad interpret oil well production logs when I was a teen, and in college. I would see these very large kicks on the log, denoting high oil content. Yet my Dad would not mark those zones for production, and I asked him why. "Look at the density side of the log, son. Tight shale. If we could ever figure out a way to produce that oil, WE WILL NEVER RUN OUT." Those last five words still ring in my ears. My Dad did not live to see this technology come to fruition, but I did.

Now, to that dollar-a-gallon gas teaser. Oil prices are set by global market forces, and any commodity is worth what someone is willing to pay for it. For decades, the Middle Eastern countries (managed by the Oil Producing and Exporting Countries [OPEC] cartel) had the upper hand. Why? Because through luck of the geologic draw, their oil is closer to the surface, and it comes out of the ground almost ready to put in your tank. Therefore, the cost of production and refining are so low, that it overcomes the cost of trans-continental transport. OPEC could manipulate supply, and the price of oil, internationally in a way that would thwart American companies from spending money on North American exploration. But modern hydraulic fracturing and horizontal drilling have changed the math. We are at a tipping point that will make domestic production cheaper than OPEC oil, under just about any scenario. At this point, we can tell the countries that hate us to (ahem) "go pound sand", and we will truly be energy independent (as we were in 2017-2020). Costs will go down, as supply goes up – maybe even down to $2 a gallon. The only thing holding back current exploration and production are restrictions by a federal government chasing green moonbeams.

The economic boom of the post-industrial world has been built on cheap energy, including coal, oil, and some alternative sources (my favorite is hydro-electric power from dams – cheap and renewable). Our modern standard of living would not be possible, nor even imaginable, without our ingenious use of nature's energy sources. And, as evidenced by our amazingly clean country (trust me, I've been to India), we can have this energy economy and high standard of living without fouling our nest.

Some people have said that we've hit "peak oil," which means that we are on the downhill side of the availability curve. Yeah, whatever. I grew up in the oilfield, and about every ten years, folks in the industry would worry that we would run out. Then some smart engineer would come up with a "new" technology to squeeze ever more oil out of those rocks. With the killer app of hydraulic fracturing and horizontal drilling, it is easy to be bullish on our energy future. All we need is the political will to produce our own energy, and the ingenuity to keep it flowing. So Dad, I raise a Crown on ice to you – your vision of an endless flow of cheap oil from west Texas is in sight!

The Miracle of "Fracking" and Debunking Myths (Part 2 of 2)

Last week's column about the "Miracle of Fracking" has led to some responses that remind me of the unfounded myths out there that people still cling to about hydraulic fracturing ("fracking"). The most popular myths are that 1) hydraulic fracturing contaminates groundwater, 2) hydraulic fracturing causes earthquakes, and 3) hydraulic fracturing somehow contaminates the surface. This column will provide the science and sense to debunk these myths, although I am sure that some of you won't buy it. I know that you folks will cling to your "fracking is evil" storyline, just like you believe that chemtrails, Bigfoot, and man-made climate change are real. There is no amount of sense that I can provide that will penetrate your incorrect beliefs, and I accept that.

Myth #1 – the "contaminates groundwater" myth. This myth is illustrated by videos that regularly circulate which show people lighting the water from their faucets, somehow claiming that hydraulic fracturing turned their tap water into an explosive menace. First off, that video is from Pennsylvania, where petroleum deposits are very near the surface. Those videos were shot years before hydraulic fracturing came to Pennsylvania, and it has been known since the first water pipes were laid in towns in PA oilfields that the very near-surface

115

petroleum and natural gas could enter poorly-sealed water mains. The reality is that every state has strict laws on how oil wells must be sealed off from shallow, water bearing formations. Regulatory agencies take pains to inspect new oil wells for concrete seal integrity.

Myth #2 – the "fracking causes earthquakes" myth. This one is actually true, but with a big *asterisk*. Hydraulic fracturing jobs recover the fluids post-process, and all that briny water has to go somewhere. Keep in mind that hydraulic fracturing water is several times more saline than ocean water. So, oil companies drill "disposal wells", generally several thousand feet deep, to depleted oil formations with available fluid space (once the oil has been taken out). The frac water is then injected through the well to its final resting place, a mile or two subsurface, never to be seen again. This process is also highly regulated and monitored. So, earthquakes? When all this water is reintroduced into thin, porous formations, it can cause minor slippage of the rock. Imagine two plates of glass laid flat. If dry, they don't move. But put some water between them, and they slide around. The good news is that these earthquakes are minor in intensity, and short-lived. The strongest ones will rattle the dishes, and maybe put a hairline crack in your drywall (if you live in the oilfield). I have a friend in Oklahoma who says that the little shakes just remind him that everyone has a job, and that his oilfield job paid for that rattled house and nice new pickup in the driveway.

Myth #3 – the "fracking contaminates the ground near oil wells" myth. This one is only true if there has been a surface spill, which is exceedingly rare. I actually managed the cleanup of a fairly large surface release of brine on a cattle ranch south of Shreveport, LA a few years ago. In that case, it was later determined that the rancher had intentionally damaged the valve on top of a disposal well, causing a 300-acre spill which killed every plant in the area. My job was to oversee the removal of thousands of cubic yards of saline soil to a landfill, which cost over $2 million. His plan was to get a settlement from the oil company for damage to his property. His claim was that a cow caused the valve to open. The judge threw out this scurrilous claim, and the guy ended up on the hook for the cleanup cost of the spill that he caused. So, unless some malfeasance is involved, the movement of spent frac water does not contaminate the surface.

I accept the unfortunate fact that some people just believe that anything related to petroleum is evil, and they will spin all kinds of yarns (yarn is made from petroleum) to demonize oil. I call these people "hypocrites", because their entire world is made possible by cheap, readily available, energy-dense, versatile petroleum. I would love to see these people live for a week without petroleum, and the plastics, heating, cooling, transportation, and inexpensive food that they deliver.

Hydraulic fracturing (combined with horizontal drilling technology) is only the latest in enhanced recovery techniques that will keep petroleum flowing for centuries. Yes, centuries. Long ago I was an exploration geologist, and many people that I grew up with in the oilfield are still employed in the industry. Just the reserves that we know about, using just today's technology, would last 500-800 years. I am confident that more centuries of supply will be added as we find new deposits, invent new ways to squeeze more oil out of the ground, and increase our usage efficiency. And, as we build more nuclear plants to serve electrical needs, we will stretch petroleum even further.

Of course, poorly-informed lefties say that we have ten years of oil left. Snort. Greenies love to lie about petroleum. So, go take a long drive in your V8, and say it with me – "FRACK ON"!!!!! Oh, and please bring the rebuttals, that you type on your plastic screen, while you wear petroleum-derived clothing, in your gas-cooled house, after you eat your breakfast which is made of food that was made cheap and available by petroleum. Oh, and enjoy our nice clean air!

The Hypocrisy of Electric Vehicles

Recently I saw a Tesla with the license plate "No Oil", and I laughed out loud. I then realized that the owner might not be making a funny joke, he might be serious. Then I reflected on the whole Electric Vehicle (EV) push by left-leaning governments. Hypocrisy is sometimes funny, sometimes sad, and always thought-provoking. So, regarding EV hypocrisy, feel free to laugh, cry, or think.

The hypocrisy of "No Oil" is glaring. EVs are MADE from oil. Tires, paint, glass, plastics, battery casings, vinyl, foam, lubricants – ALL made from petroleum. Over 50% of the energy source is from petroleum (see next paragraph). The roads they drive on are largely petroleum. A diesel-powered truck delivered the EV to the owner. A petroleum-powered tow truck will be necessary when these things break down, or the electronics just lock up. "No Oil"? EVs couldn't exist without LOTS of oil.

Oil (or coal, another hydrocarbon) is also the primary source of energy to run an EV. Wait, you say, they run on electricity, stored in a battery! I am sorry to inform you that electricity is not a SOURCE of energy, but a delivery method.

As you may remember from my Energy series, Arizona gets the following percentages of its electricity from these sources – Petroleum 44%, Coal 13%, nuclear 29%, Wind/Solar 9%, and Hydropower 5%. US averages differ slightly by location, but are close to Arizona's.

The ONLY place where EVs sort of make sense is where the population density is very high, and air quality is poor. EVs simply move the emissions out to power plants, away from the city, limiting tailpipe emissions where they are a minor problem. The places that meet these criteria in the US can be counted on one hand – LA, San Francisco, Seattle, Denver, and New York City. Literally everywhere else does not have this problem, and EVs "solve" a problem that largely doesn't exist.

Of course, greenies will scream and cry about "saving the planet", but Earth is doing just fine with or without EVs. In fact, the strip mining for battery metals, mostly in China or Africa, is an environmental nightmare. Metals strip mines leave a scar forever, damage water and wildlife, and create horrific child labor situations. When an oil or gas well is spent, the pipe is sealed below ground, and no surface expression remains.

Then there is the hypocrisy of cost. Lefty governments push very hard for everyone to buy an EV. I asked my server at breakfast one morning if she was in the market for an EV. She laughed, and said that her 2006 Corolla is her only choice, financially. Yet lefty governments, including some states, are demanding that only EVs be available for sale in 10-12 years. Like everything EV, the math just doesn't work.

Three limiting factors will come into play, to keep EVs as the niche vehicle they are (the niche being urban, wealthy lefties with $70,000+ to spend, who never tow anything, don't take trips, and have the time to wait for charging). LIMITER #1 is battery metals availability. Our anti-mining laws limit mining in the US, so we rely on countries who hate us (mainly China). Not enough battery metals are or will be available in five years to build even 15% of the fleet worldwide. LIMITER #2 is charging capacity. Our grid can't take much more drain (evidence California brownouts in summer), and power companies don't have the money to upgrade the grid sufficiently (unless you want your power bill to quadruple). We will hit the grid limit at about 10% of the fleet as EVs, if we even get that high. LIMITER #3 is demand. Studies show that only 25% of buyers would even consider an EV (a lesser number have the money to buy one). And more importantly, 41% say they would never buy one. Of the 58.1 million vehicles sold worldwide in 2022, about 1.9 million were EVs. So, EVs account for roughly 3 of every 100 new vehicles sold globally (and are under 1% of existing on-the-road vehicles).

EVs will never be greater than 10% of the US new-vehicle fleet, slightly more in Europe (with their incredibly high fuel taxes), and China (where a huge new dirty coal power plant comes online every ten days). When battery metals either run out, or become prohibitively expensive, owners won't be able to replace the batteries, and their expensive EVs will become stationary yard art, worthless. In the worldwide vehicle market, EVs will never even get a toehold in the 70% of countries that do not have a reliable power grid, are perpetually poor, or have common sense.

Finally, why do governments push EVs? To feed a green fever dream, when we don't have a serious air pollution problem? No, it is a matter of control. With the flip of a switch, the government can shut off charging for any reason, such as grid overload, "health crisis", or they don't like your social media post. The electric utility business is already highly regulated, with government tentacles all through it. Big Brother lives in our electrical system.

So, thanks but no thanks, I'm not interested in buying a hypocritical, limited-function, inconvenient, expensive golf cart. No, I will keep my internal combustion engines, with their efficiency, versatility, ease of fueling, virtually unlimited range, and measure of privacy from government control. In fact, I believe I will go for a nice long Sunday drive in the mountains today, and enjoy the blue skies, clean air, clean water, and plentiful wildlife! See you on (or off!) the road!

"Our Stressed Electrical Grid"

It sits quietly, out of sight, taken for granted, generally unnoticed. Then when it fails, society is thrown into chaos within hours. The "juice", power, our ubiquitous electrical web that we all rely on for a myriad of devices and comforts. Our power grid is not a sexy topic, it lurks in the background. Perhaps you have heard snippets here or there about our power grid, and that it is in peril. Then you went about your day, forgetting about the issue in moments, because you want to microwave some popcorn and watch Seinfeld reruns in the comfort of your air-conditioned home. Well, prepare for a little eye opener. Our grid is vulnerable in many ways, and we aren't adequately addressing the vulnerabilities.

Our power grid's vulnerabilities fall into three broad categories – 1) external threats, 2) internal threats, and 3) systemic threats. Each of these threats are serious, and deserve attention. Unfortunately, that "attention" translates into money to fix it. LOTS of money. And I don't think anyone is excited about a power bill that is 4 to 10 times your current bill. Got your attention yet?

EXTERNAL THREATS include cyberattacks and direct attacks on equipment. CYBERATTACKS are basically "hacking" by an outside entity, into the computer systems that run our grid. This has already happened several times, in the form of shutting things down remotely just to cause mayhem, or shutting things down using "ransomware" where the perpetrator holds the grid "hostage" for monetary payment. And yes, it has worked. DIRECT ATTACKS on equipment are basically vandalism, targeted to unguarded equipment like transformers (almost all electrical infrastructure is unguarded). These attacks have been carried out by malicious vandals, as well as organized groups who want to gain attention for their "cause". Some have used attacks on remote electrical equipment to start forest fires, which is both a grid attack and arson.

INTERNAL THREATS include AGE of the system, and the insipid threat of increased DEMAND. AGE of the system is a bigger deal than you think. Many transformers, step-down stations, power lines, and other hard infrastructure are in use well beyond their design life. Economically limited power companies find that if a transformer is still working, they will often let it keep going until it fails, rather than performing proactive replacement as equipment ages out. This results in weaknesses, and power failures for simple reasons like parts wearing out. Proactive replacement is effective in minimizing this threat, but it costs more. DEMAND is an ever-increasing threat. With our growing population, migration from high tax states to the Sunbelt, and more recently letting millions of illegal aliens just walk across the border, the demand for power, especially in the Sunbelt states, is increasing. Another unnecessary cause of demand spikes is the government's ridiculous push into electric vehicles, which adds enormous demand to the grid. If the government's EV projections (and mandates) even come half true, nobody knows where all this excess power will come from, and more blackouts will ensue.

SYSTEMIC THREATS come from our enemies, and nature, both of which we cannot control. EMPs (electromagnetic pulse devices) are weapons that could be exploded in our atmosphere by any number of enemies (China and North Korea brag about this capability). The electromagnetic pulse is like a nuclear explosion that affects electronics. Basically anything with a micro-circuitry chip that is located within hundreds of miles of the blast would be ruined, by frying the micro-circuitry chips that make the modern world go. Welcome to the 17th Century. The really big threat that we can't control is from our Sun. Solar flares are intense bursts of radiation (solar storms) that our sun throws out periodically. Most of them miss the Earth, but if we get a direct hit, satellites, electrical systems, and the internet would go down, possibly for years.

Some of these threats can be mitigated with giant gobs of money, for example protective systems against cyberattack, circuit interrupters for pulses, or just building more power plants and replacing distribution equipment. Some threats cannot be mitigated, like an EMP attack, or if the Sun decides to throw a big flare our way.

So, choose the source of your gray hairs. You can worry about the mythical "Man-made climate change" (with no demonstrable effects), or what Chemtrails are doing to you (snort), or a Bigfoot attack. Or you can worry about the very real threat presented to our way of life that is a fragile power grid. To learn more, visit GriddownPowerup.com for lots of resources, and a link to a documentary about the topic. Meantime, learn to play a musical instrument, because when the power goes out for years, you are going to want some entertainment. Oh, and things like food, too. When the grid goes down for months, that means no large scale refrigeration, no fuel pumps for cars and trucks, and no stock for grocery stores.

Sorry to be a Debbie Downer here, but our power grid is a very serious Achilles heel to our modern lives. Right now, our defense is pretty much limited to "gee, I hope none of this happens". Hope is not a very effective mitigation strategy.

Oh, and a shout out to the power company employees, linemen, and others who keep the juice coming. Talk about a thankless job! But count my voice as a big THANK YOU to power workers. You are indeed appreciated, and let this be a jolt to your self-esteem (see what I did there)?

CHAPTER 4

Water Topics

INTRODUCTION

Water, as used in this chapter, refers to groundwater and its availability in the mountain basin where we live. This chapter is specific to a certain area of Arizona, but corollaries may be drawn to any number of locales in the arid Western US. The impetus for this three part series was my attendance at a forum about water issues in our county, and I was astonished at the general lack of understanding of groundwater by the public.

Most people apparently think that the water beneath our feet (if we are lucky enough to have it, groundwater is NOT everywhere), is just a big lake or river underground. The first column in this three-part series endeavors to educate people about what groundwater is (and isn't). The second column goes into the incredibly complex topic of Western Water Law, and believe me, getting even the basics into a 900-word limit was a challenge. The third column covers how groundwater can get ruined and nobody can use it, man or beast.

This topic is a great example of how the reader can do a lot of internet searching on his/her own, to answer the myriad questions that come up when thinking about groundwater as a water source. If you live in Arizona, I suggest you take any chance you get to hear Grady Gammage Jr. speak. He is a water lawyer, extremely knowledgeable, and a wonderful, understandable speaker on the topic.

Groundwater – Defined

This column is the first of three about a topic that generates endless controversy in the desert, groundwater. After listening to arguments about groundwater for my whole life, I thought it might be useful to apply a little "Science and Sense" to the discourse. The average arguer on the topic would get about a 33/100 for understanding what groundwater is, how it is shared, and how it can get ruined. This first column focuses on what groundwater IS.

A good place to start is to dispel a common misunderstanding. Groundwater is not a big lake or river underground. It is water that exists in the pore spaces in rocks, usually sandstone, limestone, loose soil/gravel, or cracks in volcanic rock. Sometimes groundwater is very deep. Other times it is shallow. Wells to extract groundwater get more expensive to drill, and to operate pumps, the deeper they go. Groundwater sometimes is "sweet," meaning that it has few or benign dissolved rock elements. Other groundwater is not so palatable, containing too much of a certain mineral element, salt, or any number of other trace elements. Most groundwater can be treated to be usable, but that means added equipment and costs. And if your well is salty, plug it. It is generally useless to plants and animals (humans are animals, too).

What we humans care about is our ability to access and use groundwater, while not over-depleting the resource. Unfortunately, occurrence and access are not uniform. An example would be useful. I have a friend on the edge of Chino Valley with a 120 ft deep well that yields about 20 gallons per minute (gpm) (very consistently) of sweet, usable water, right out of the pipe. His neighbor's well is about two miles away as the crow flies. It is 150 feet deep, yields about 5 gallons per minute, with production rates from 2 gpm to 15 gpm. His water is not so tasty, and is laden with calcium. This water requires treatment, and his well screens, pipes, and pumps plug up and wear out more often. Both wells are part of the monitoring program administered by the Arizona Department of Water Resources (ADWR), and they are monitored regularly for water levels. My friend's well doesn't vary by more than ten feet per year in depth to water (regardless of precipitation) – his neighbor's swings within about a 50-foot depth range.

This situation illustrates the problem with "predicting" groundwater. Rather than the uniform "lake," or even a uniform "water table" (a terribly misleading term that should be avoided), groundwater is an extremely complex series of interconnected sub-basins. Imagine if you will, a large valley. Looking across it, imagine every square inch of that valley is covered with coffee cups, swimming pools, horse throughs, cereal bowls, and hot tubs. Some of these vessels are connected to every vessel around it, some are only connected to one or two, and some leak. And the connections themselves aren't uniform – some are soda straws, some are garden hoses, some are 8-inch diameter pipes. This huge field of vessels, of varying size, on a slope, with varying connections and efficiency of connections, is pretty much how groundwater eventually works its way from the high points in the valley to the low point. And

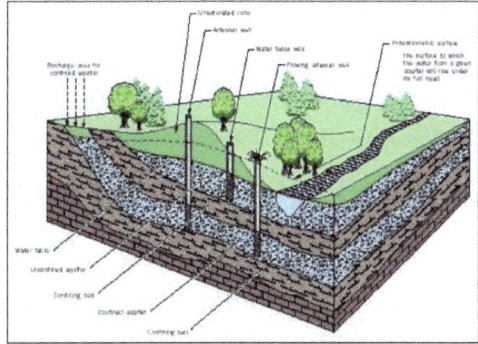

water only flows downhill, even though a water law attorney I know swears that water can run uphill, if it is toward money.

Now that you have this picture in your mind of a valley full of vessels, imagine that the bowls, troughs, and swimming pools only represent one level of groundwater. With the layering of geologic units (think of the wall of the Grand Canyon), a "groundwater basin" may have three, five, or 17 such interconnected layers. And some of the layers leak down to lower layers.

All groundwater comes from rain or snow. In the highest zones, nearest the surface, annual rainfall has the greatest effect on water levels. Droughts affect these shallow layers more directly. Deeper layers contain water that may have fallen from the sky hundreds or thousands of years ago, and they also gain water that percolates down from above, where the separating rock layers "leak."

131

So, now that we understand the basics of where groundwater comes from, and how it moves, think about my friend's well, and his neighbor's. My friend's well taps into a swimming pool, one that is well connected to the other sub-basins around it, and can be replenished easily for constant draw. His neighbor's well may tap into a cereal bowl, poorly connected to the vessels around it, in an area where more calcium-rich rock is located. Two miles difference as the crow flies, but lots of difference in the water resource.

Next week we will cover how the ADWR looks at the availability issue, while carrying out their role of deciding who can drill a well, how much can be drawn, and how man, animals, a river, and any other number of users can SHARE this limited resource. When I talked to a friend who recently retired from ADWR regarding the gnashing of teeth about groundwater in our area, he just laughs. Our area is located atop the richest groundwater basin in the state. Now that is like saying we are the richest guy in the trailer park, because we are still located in a desert. But he assures me that ADWR is not worried about groundwater availability in our area. The state has much more critical issues in other basins, which aren't as rich as ours. Despite what activist groups with an agenda say about it. But that is a teaser for the next column in this series.

Groundwater – Who gets it?

This second column of three about groundwater will cover how, and who, decides who gets the water. Last week's column set off the expected wailing from the local activist group who thinks they "own" the topic, but the reality is that understanding groundwater systems is different from pushing an agenda by spinning data. Activist groups, some who want to "stop growth," and others who want to "protect the river," are very good at spinning the science to tell their biased story. But here, Science and Sense attempts to tell the straight story, with no agenda, and to do it in three 900-word columns.

Last week's column, explaining the basics of groundwater, shouldn't be controversial. Facts are intractable things. Where the shouting starts is when the discussion turns to who GETS the water from the ground. Western water law is a convoluted mess, the result of 150 years of legal fights about the most essential resource in the desert, water. I couldn't do the topic of western water law justice in fifteen 900-word columns. Most of the shouting is about allocation of the Colorado River, but in Yavapai County, we are unencumbered with fights about Colorado River water, the Central Arizona Project (CAP and its canals), the Colorado River Compact, and a dozen other topics that most of Arizona DOES fight about. We don't get Colorado River water, basically because it would be wildly expensive to pump it a mile uphill. We sold our CAP water rights to valley municipalities years ago, which was a good deal for everyone (since we would never be able to get that water up the hill anyway). So, we live within our groundwater budget.

Since our area is adjudicated by the Arizona Department of Water Resources (ADWR) as an "Active Management Area" (AMA), it is the state who has the decision-making power over the way our groundwater is accessed and used. ADWR's charge is to adjudicate water uses to the benefit of multiple users – farmers, municipalities, private property owners, the river, and the silent 800-pound gorilla, the Salt River Project (SRP). Yes, SRP operates reservoirs along the Verde River (Horseshoe and Bartlett Reservoirs). These lakes are part of a system (with five other major reservoirs on the Salt River) that provides power and water to lowlands in Arizona, where the majority of people, politicians, and money reside.

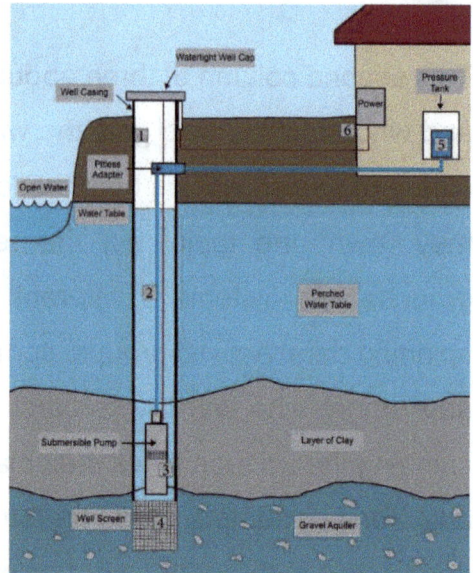

Whenever a water user wants to expand their draw on the aquifer (or add a new user), they must get a permit from ADWR. ADWR reviews lots of technical data, well information, and known aquifer information as they consider the permit. BALANCE is a big part of their consideration process. Existing water rights and uses, preservation of a certain amount of base flow for the health of the river, and saving water for future municipal users all play into ADWR's consideration.

ADWR uses computer models as part of their technical analysis. This author has used these models for years (in the context of contaminant movement), and is very familiar with the way variables can be manipulated toward a desired result. Activist groups sometimes fight permit applications, by providing their own models with variables manipulated toward their desired result (showing less available water). But make no mistake, ADWR is the final decision-maker on groundwater use permits. Applicants skew their variables to show plenty of available water. Activists skew their variables to make dire predictions that man's uses will drain the aquifer. ADWR has been playing this game with applicants and activist groups for decades, and is very good at sifting through the BS to get to a reasonable approximation of the truth.

But as we learned in last week's column, there can never be enough data from wells to provide 100% accuracy, given the complexity of groundwater systems. It is all an educated guess. Right now one side is screaming that the groundwater resources in the northern end of our basin shouldn't be touched, because the river will run dry, animals will die, then humans, then we become California's Owens Valley (look it up). The reality is that the northern end of our groundwater basin appears to be quite rich in available water. Some of it can (and will) be withdrawn to serve population growth, which is coming (like it or not). ADWR will determine whether enough water can be drawn, without damaging the river, to supply 10,000 new residents, 100,000 new residents, or 200,000 new residents. Pick your number, the lottery draw is in about 30 years.

And the new neighbors ARE coming. Probably not 100,000 of them in our lifetimes, but population will increase. Trying to block groundwater access through fear is akin to trying to go on a diet by burning down one cornfield in Iowa. The food will keep coming to the dieter. People will keep coming to Yavapai County.

And recent efforts to classify the Upper Verde as a federally-protected "Wild and Scenic River" are an attempt to take some of that groundwater "off-limits." Now, does the Upper Verde deserve protected status? This author believes YES, but opinions vary. But the amount of water that will be "protected" is a drop in the bucket (see what I did there?) relative to the amount needed to support municipal growth. ADWR will make that call, years (likely decades) from now.

People will continue to fight about water (it is what Westerners DO). Activist groups will go about their useless fear-mongering and political posturing (and they STILL don't make the decision). But keep in mind the loudest, silent voice in the fight. SRP, with their friends at the legislature, and the unstoppable need for electrical power and water down in Phoenix. If you're betting on horses here, the odds are on SRP.

Groundwater – How to Ruin It

The first two columns in this three-part series covered a definition and basic understanding of groundwater, and the way people fight about accessing this liquid gold of the desert. As any discussion of water will do in the desert, the first two columns made some people mad, and they are yelling now, can you hear them? This was completely expected – heck, it is easy to take a roomful of people, and go from friendly discussion to a bar fight in minutes. Just holler "GROUNDWATER."

This column will cover how groundwater can get ruined, so that it isn't a benefit to anyone. Having spent a career in soil and groundwater cleanup from contamination, this author has seen firsthand how we can locally ruin a groundwater resource. Fortunately, we have gotten much better in the past few decades about protecting our Earthly home, but natural and man-made processes can still render groundwater unusable on a localized basis.

Most of my career was spent dealing with the aftermath of leaking underground storage tanks, and improper disposal of liquid and solid hazardous wastes. The largest, most distributed, and most costly offender (for decades) was the common gas station. From the 1920s until now, tens of thousands of gas tanks beneath stations leaked. Part of it was poor tank construction, part of it was corrosion, and some of it was wanton disregard for disposal of waste products. I have personally overseen the excavation of hundreds of old gas, diesel, and waste oil tanks, and my experience mirrors what the EPA reports – that about 80% of all underground storage tanks ever deployed, leak. But the silver lining is that gasoline and diesel, in most geologic settings, will slowly self-remediate. This is because hydrocarbons oxidize, and nature provides microbes and chemical reactions that will, over time, break down gas and diesel. Of course, if the stuff gets into your well, or a municipal well, it is still a short term, localized problem.

Far longer lasting an insult to groundwater comes from certain industrial polluters. Dry cleaners used to dump their spent fluids (carbon tetrachloride in the early days, then perchlorethylene ("perc")), in the unpaved alley. These chemicals don't float atop groundwater like gas or diesel - they are heavier than water, so they sink. This makes finding and cleaning up these chemicals much harder. In the cleanup business, I used to hear "a perc spill is forever." A similar chemical class includes degreasing fluids used in micro-circuitry plants, which were built in Arizona from the 1950s to the 1980s. Look up the "52nd Street Motorola Plant" in Phoenix for a good (bad) example of a very large, and very long-lived, contamination "plume." The West Van Buren plume is also miles long, and while the source is always referred to as "unnamed commercial sources", a quick look at a map will show that the epicenter of the source area was Mission Linen, on 7th Avenue just north of Van Buren, which has operated for decades. And large industrial polluters (in cities that were the center of industrial production in the mid-20th century, think Detroit, Houston, or LA), left a smorgasbord of toxic legacies underground.

Groundwater Pollution

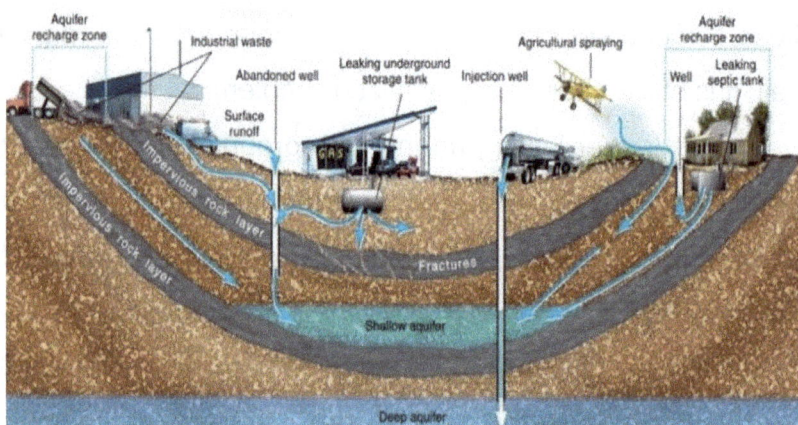

Some groundwater contamination is natural, and is usually only a problem if we put wells into it. Early in my career I worked in South Florida (Ft. Lauderdale / Miami). There, municipal wellfields pull enormous amounts of fresh water from the subsurface, which draws in subsurface salt water from the beaches miles away. Once a municipal well experiences "salt water intrusion," the only recourse is to abandon the well, and drill further inland.

Hard rock mines can also negatively impact groundwater in the vicinity of the mine. Just look at Colorado, and all of the damage from acid mine drainage. I go high-altitude off-roading every year in the San Juan Mountains between Durango and Ouray. That area is a master class in how to destroy watersheds through poor mining practices. Of course, all of those mines operated over 100 years ago, when nobody was thinking about the creek. They were only thinking about the riches in gold, silver, lead, and molybdenum that could be dug from the mountains.

More recently, the "forever chemicals" of PFAs/PFOSs have come to the fore. Perfluoro-octanoic Acids (PFAs) and Perfluoro-octyl Sulfonates (PFOSs) are ubiquitous chemicals found in everything from firefighting foam to the coatings on plastic food containers. The problems with these chemicals are many – we don't yet know what concentration is toxic in humans, these chemicals are found pretty much EVERYWHERE, and (currently) no known method exists to effectively remediate groundwater impacted by these compounds. Even investigating PFAs/PFOSs is challenging, because environmental sampling equipment is made from the very plastics that often contain these compounds.

Any number of other things can render our groundwater contaminated. Naturally occurring radioactive rock. Pesticide/herbicide runoff from farming operations. Nitrates from feedlot operations (what a crappy subject). Improper functioning of septic systems. Residue from Chemtrails, alien

141

aircraft, Bigfoot scat, or Jimmy Hoffa's gravesite (LOL just joking, seeing if you're paying attention).

The bottom line about groundwater is to not take it for granted. Groundwater accounts for 1.69% of all the fresh water on Earth. Half of that is inaccessible. Half of THAT is suitable for human consumption. So, about 1/3 of 1% of the water on Earth is accessible, fresh, usable groundwater. That is a lot of math for a Sunday morning, but know this: every drop of our groundwater is precious. Please treat it accordingly.

CHAPTER 5

Community and Growth Topics

INTRODUCTION

Growth is occurring all over the world, as the population continues to soar. It is especially felt in small to medium sized towns (like Prescott and Prescott Valley, total combined population about 95,000) where many newcomers have relocated here from large cities like Phoenix or Los Angeles. These folks do NOT want their idyllic mountain town to become Orange County, CA. When I ask people when they thought the town "got too big", the answer is consistent, whether they arrived six months ago, or in 1996. People generally report that the town got too big within a year or two of their arrival. Hmmm.

A couple of columns about this author's experience with an overreaching and inept HOA follow. I am happy to report that the community rose up, and voted down some overreaching rules that the Karens wanted to impose upon us. Fighting back at the ballot box sometimes works!

Other community-minded columns in this chapter include the topic of citizenship, the benefits of a great Farmers' Market, being smart about where you buy your home, and the perils of driving in a town with lots of "well-seasoned" drivers.

I use Prescott as the topic of many of these columns, but I could name dozens of other towns that deal with the same issues. Prescott's motto is "Everybody's Home Town", and lately, lots more people are taking that literally. No matter where you live, growth and community issues aren't going away, so let's all just get along!

The Science of Growing Our Town

Recently I have been involved in discussions about the growth of the Yavaplex (Prescott, Prescott Valley, Chino Valley, and Dewey/Humboldt) and how people perceive and feel about it. Related to this topic, but also separate topics, are roads, schools, taxes, and water – each with a spot on the "growth" spectrum of topics. Everyone seems to have an opinion, and the opinions vary widely.

First, the basics of growth. We are a wonderful collection of communities in a beautiful foothill setting, with history, culture, touchstone events (Rodeo, Christmas, Cowboy Poets, etc), and balanced, mild weather. Therefore, people are attracted to the area as a great place to live. This is GOOD. However, high-paying jobs are not as plentiful as larger cities, so we tend to attract a higher percentage of people who don't need those jobs, or have retired from them. We have elevated housing costs, because the housing stock is generally higher end, to serve the needs of those well-off retirees. This leads to a shortage of workers for traditionally lower-paying jobs (evidence the Help Wanted signs everywhere you look). This labor imbalance is BAD.

Second, people have a wide range of opinions on where, how much, and how fast the area is growing its population. On a social media thread, I asked people three questions: 1) When did you arrive, 2) What was the population then, and 3) What do you think is the ideal population of say, Prescott? The purpose of these questions was to get people to think about their role in growth, and whether they thought they were part of the problem or not. Most people didn't answer the questions, stating simply "we're too big" or "Prescott is already ruined". A few did answer the questions, and invariably, they thought that the town was the right size when they moved here, and got too big soon afterwards (therefore not counting themselves as part of the "overpopulation"). This too, made sense – why would anyone move to where they thought it was overpopulated already?

Third, perceptions of growth had a wide range of measures. One respondent said that Prescott got too big the day the second Wal-Mart opened (a commerce metric). One lifelong resident said that about 1989 was the tipping point (a linear time metric). Still another responded that when it took longer than 15 minutes to cross town, Prescott had gotten too big for its britches (a transportation metric). Yet another complained that when it got too hard to park on the courthouse square on a weekend, that was the sign we were overpopulated. So, perceptions are individual, and it was telling that very few people would honestly answer about whether they thought they were part of the population "problem" (which, by definition, they are).

A recurring theme was a general dislike of the newer developments, with postage-stamp sized lots and cracker-box houses with no personality. One person called them "Rooftop Orchards". The new developments of this style are indeed jarring visually, bereft of trees (or even room to plant them) and without yards, personal space, and room for kids to play. And developers build this way (factory-style) to get the maximum bang for the buck, selling the most houses in the smallest space. In a tight housing market, people buy these boxes (although I can't imagine why, unless they are really cheap). Until people vote with their wallets, and stop buying these little rabbit warren homes, the developers will keep building them.

Some cried "not enough water", a refrain we have heard from a local activist group here for decades. Water fear has never been shown to be an effective growth deterrent itself, but the goal is to get enough people terrified that the tap won't turn on, so that they will vote for elected representatives that will try to curtail the evil "growth". I have seen this fear-and-manipulation tactic many times in the arid west, and it doesn't work. People's desire to live in a certain area, and market forces, will overcome water fear every single time. And an inconvenient truth is that Yavapai County is relatively rich in groundwater resources, compared to other rapidly growing areas of Arizona (not counting Phoenix and Tucson, with their multiple water sources).

Roads are another pinch point when it comes to a growing population. Roads will become noticeably more crowded with growth, and the cycle of roadway improvements, widening, and adding new routes (like the SunDog Connector), operates much slower than the growth curve. Anyone who lived through Phoenix's growing pains from gridded arterial streets to a network of freeways can attest to the lag between need and solution.

Bottom line, growth is coming, like it or not. You can scream at the sky about it all you want, but the best we can do is to plan wisely for subdivisions, roads, and water resources, so that the growth is well managed. One respondent to my social media query had a great point about growth, versus the alternative. He came from a town in Wyoming that was dying. Job losses, dropping property values, emptying schools, loss of shopping choices. I don't think Prescott will have to face that eventuality anytime soon, so we should make the best of what is happening. Attitude is a big part of it. Get involved, welcome your new neighbors, and roll up your sleeves to help build the kind of town you want to live in. After all, it is our home!

One parting shot – not all Californians are evil. Many come here as refugees from political madness, and chose Prescott purposefully. They care deeply about preserving the charm of this town and the values it represents. I say, give them a chance!

The Science of Growth (Revisited)

Last year, I wrote a column about the growth of our Mayberry-in-the-Pines. You can use the search function, and go back and read it again – I don't plan to cover the same ground. In the last six months, I have received lots of comments, information, studies, and a few spirited discussions at certain road project open houses. So this time, I want to cover a few more fine points about the inevitable growth of the Yavaplex.

I reviewed several studies about population growth, planning processes, and water availability. I will start with water, since that seems to be a sticking point with many no-growthers. The water studies agree that we have plenty of groundwater. Whether it is "paper water" or "wet water", the Arizona Department of Water Resources (ADWR) and information from actual wells agree that we have ample water for at least double the current population of the County (even in drought). We have never exceeded the amount of water that ADWR allocates for City use, and in fact, our usage per resident is actually DOWN, through smart landscaping choices and voluntary conservation (although we could still improve). The studies (and ADWR opinion) are clear - ample water exists, it is just a matter of getting it from where it occurs, to where it is needed (the municipal part of usage, anyway). And every day we wait on accessing Big Chino water and building the pipeline, the less affordable it gets. Finally, remember that many of those who look out the window at a desert state and scream "There's No Water"!!! are operating on manipulated emotions, not facts. Oh, and the Upper Verde has remained wet, through 26 years of drought.

As for growth, the city of Prescott has been growing at a consistent rate of about 2% per year, for many years. There is no "rush" or "stampede". Other Yavaplex towns might be growing slower or faster, but the growth rate is effectively moderated by two primary factors – 1) availability of housing stock (affordability plays into that), and 2) jobs. I laugh when the no-growthers scream that we are becoming Orange County (an unhinged person on social media actually accused me of advocating for that). The reality is that Yavapai County will NEVER be a densely-developed megaplex. We have too many land use restrictions, and not enough jobs, to ever approach that level of density. Other factors limiting our growth are our relative remoteness (not located on an Interstate, two hours from either Phoenix or Flagstaff), lack of a major airport, and our inadequate health care scene. Don't get me wrong, people will still move here because it is beautiful, but those factors will continue to limit growth.

Having worked for a transportation engineering firm for decades, one comment that I hear still surprises me - "just don't build more roads". Ask anyone who saw what happened to Denver in the 1980s/90s, or Austin in the 1990s/2000s, what happens when you try to discourage growth by keeping your roads small. Gridlock, soaring housing costs for close-in homes, and increasing road rage. These same folks love to say "just slow down, enjoy the view". Aw, thanks Aunt Bee. But when you are forced to slow to a crawl, and it takes 45 minutes to get across our small town, attitudes might change. We already see an increase in accidents, rudeness, and frustration because of a road network that hasn't kept up, exacerbated by too-slow and inconsistent speed limits, and inattentive drivers. Aunt Bee rapidly becomes Ernest T. Bass.

You don't have to travel far to see the result of the "no growth, small roads, stop building" mindset. Sedona is unaffordable, hard to get around, and has lots of angry drivers. Over twenty years ago, the Village of Oak Creek built a dozen tiny roundabouts, primarily to make it miserable for Phoenicians to get to Sedona (in an attempt to force them around through Cottonwood). Then Cottonwood built a bunch of roundabouts-in-the-middle-of-nowhere, seeing how effective these roadblocks had been in the Village. Now both towns have to live with these traffic-clogging abominations.

Here is the hard fact - YOU CAN'T STOP GROWTH. The best you can do is plan wisely for it. We must get ahead of the gridlock with additional lanes, timed lights, and additional new roads where we can. We must plan ahead for increased water usage, commensurate with the steadily growing population.

The only alternative is to turn into Sedona, Denver, or Austin, and nobody wants that.

A final thought about this growth debate. I have encountered some really nasty no-growth people. They absolutely HATE newcomers, or anyone who tries to offer common-sense solutions. "Go back where you came from" is a common epithet, along with "Well, how long have you lived here"? Sorry, venom-spitters - the minute that a newcomer gets a house (bought or rented), gets a driver's license, and starts paying taxes here, they are just as much a Prescottonian as someone whose ancestors moved here in 1901. And I ask you, what was on your lot before YOUR house was built? We are all part of the growth.

Also, let me be clear: I am NOT "in the pocket of developers" (LOL), and I am NOT unrealistic about water (sorry, I follow facts, not fear-mongering). Meantime, let's focus on realities and solutions, limit the emotional outbursts, and try working together to provide adequate roads and water for the future. Until we do that, the growth discussion will just get uglier.

Of course, you can always move to Seligman. . .

The Science of Inept HOAs

Once upon a time, there was a neighborhood with a Homeowners Association (HOA) – let's call it Mildweed. The neighborhood was well-established, the properties well-maintained, and the neighbors (generally) got along. The neighborhood really didn't need an HOA, because of the attributes of the area and its residents, but it existed anyway, not really doing much of anything. And it was cheap – under $200 per year. The one shared asset of the neighborhood was a community pool (a smallish backyard-style pool, not a water-park extravaganza), which was nice for the 30 or so households (out of 200) that used it. This pool was easily maintained by the $30,000 or so raked in annually by the HOA dues. The Mildweed HOA had saved up over $70,000.

Then, in 2020, the "Pool Committee" started rattling its sabre that there was an "urgent need" for pool repairs. The few residents who paid attention to this alarm pointed out that the pool was fine, including the deck, tile, railing, and fence. Nevertheless, this "urgent need" warnings slid by for three more years, with the pool functioning as intended, the same 30 or so families using it, and the rest of the 200 lots dutifully paying their HOA fees and building up the bank account.

Then 2023 comes along, and the HOA Board springs another "urgent pool repair" warning at an October meeting when many of the residents had returned to the Valley (a significant number of homes in Mildweed are seasonal homes). With a minority of residents in attendance at the meeting, and with the importance of the meeting kept under wraps, the Board sprung a whopper. Apparently, the ONE BID that they received for the pool repairs "urgently" needed to be voted on and accepted, for $90,000!!!!! Yes, you read that right, $90,000 for pool repairs!

Of course, the $90,000 wasn't just for pool repairs. It included actual "repairs", like decking, tile, railing, and a heater (still not sure why the pool needs a heater). But the scope of work also included some questionable capital improvements, like a new, border-worthy "security fence" and an electronic surveillance system. Really? For little Mildweed? The residents hadn't been reporting hordes of marauding swimmers, and frankly, the pool was only used for a few hours a day, during the 3 months of summer.

There were a few people at the meeting who raised valid points, about the impropriety of acting on only one (ridiculously high) bid, including unnecessary items, and waiting three years from the time that the situation was deemed "urgent". The HOA board, knowing that they would likely lose the vote with those present at the meeting, CHANGED THE VOTING RULES at the meeting, and decided to send out a ballot to the membership.

Now, voting is a good thing, when conducted properly. But the ballots were sent out hastily, and only offered one option, yes or no, to take the ridiculous $90,000 bid, or leave it. No option to only fix what needed to be fixed, no option to consider decommissioning of the pool and reuse of the area for another (year-round) purpose, no presentation of the ONE bid for inspection. The ballot was also sent out with a short voting response time, during a month when many residents weren't even there to receive the ballot. In addition, the ballot was worded in such a way that the ENTIRE scope of work was all totally necessary, hiding the fact that much of the scope was unnecessary. Oh, and the ballot included an assessment to every lot of $300, raising $60,000 toward the effort. Remember, the HOA already had over $70,000 sitting in the bank collecting dust.

A few short weeks later, another mailer was sent out by the bully board – Surprise, the pool repair vote "overwhelmingly" passed. 78 voted yes, 38 voted no, and tellingly, 82 did not vote. It is hard to see how 78 of 200 makes for an "overwhelming majority". But the bylaws were written so that the majority of "votes cast" rule the day, not the majority of people that have to pay the assessment. So by sending out a ballot with a biased description, with no options, during a time of year when less residents would even see the ballot, tilted the scales in favor of the $90,000 boondoggle.

Oh, and did I mention that the Mildweed HOA already has over $70,000 in the bank? It seems like they could have waited for more bids, and gotten a better price (spending the saved-up

156

dues, and negating the need for a "special assessment"). They could have offered a ballot showing ALL the bids, with the needed repairs separated out from the capital improvements. They could have done a better job of outreach, to get most of those 82 non-votes to vote. It seems like they could have gone door-to-door (the neighborhood isn't that big) and presented the issue directly, and given EVERY resident the opportunity to consider the issue.

But no. It was easier to be sneaky (an October vote), to be lazy (moving forward after only one bid), to be duplicitous (adding unnecessary items and passing them off as "necessary"), and to be opaque rather than transparent in the process (questions are so annoying). Is any of this familiar? Maybe they should have called the vote an "Omnibus".

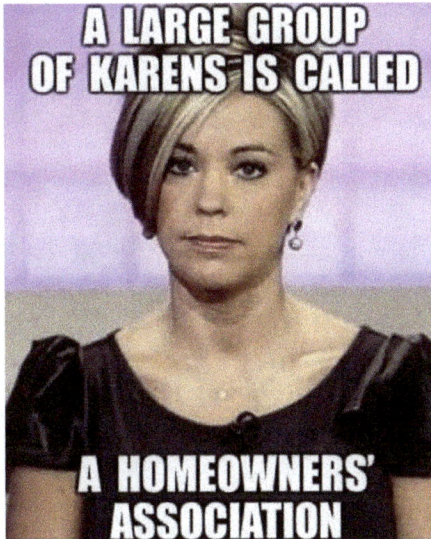

A LARGE GROUP OF KARENS IS CALLED

A HOMEOWNERS' ASSOCIATION

Maybe this Board was all of those things – sneaky, lazy, duplicitous, opaque. Or maybe it was simple ineptitude. You be the judge – or wait until an actual judge gets a say in this (stay tuned). If you live in an HOA, keep a close watch. Make the Board accountable. Because if you don't, you might find

157

yourself in this situation, too. Feel free to share in the Rants and Raves your thoughts on HOAs, they tend to have a reputation, and I wonder if the Mildweed board is typical, or is an outlier.

The Science of Inept HOAs, Part 2

Some of you might recall my column from last fall, where I introduced the shenanigans of a hypothetical local HOA, the Mildweed HOA. That column outlined an illegal voting process, to sneak through an "urgent pool repair" need, for a bid of $90,000. The pool "committee" (one person) acted upon one bid, clearly a "we don't want the job" bid. The Mildweed HOA also never came to the community to ask whether residents thought the pool was worth keeping, or if a lesser scope could achieve the objective. The pool is only used four months per year, by less than 20% of the residents, and is a significant upkeep and insurance expense. The Board jammed through an illegal "special assessment" (not allowed in the current bylaws), and slapped everyone with a $300 bill. Oh, and they never offered the ballots for a third-party review.

Now, 10 months later, the pool work is mostly complete. On prodding by residents, the Mildweed HOA finally released an accounting of what they spent, and it was a whopper - over $100k. For pool repairs (!!??!?). This is what happens when one person, with no other input, makes a decision using only one bidder who clearly didn't want the job. And that incompetence has cost everyone in the neighborhood.

And now the Mildweed HOA Board is at it again. In advance of the annual meeting on Oct 6, they have put out a set

159

of "revised" bylaws and CCRs (the rules they enforce upon residents). The Board never sent out a questionnaire to the neighborhood regarding what we care about today, they simply revised a 50-year old set that has been revised and butchered through the years, always increasing HOA "powers". This mess they sent out not only contradicts itself (and in some cases goes afoul of state law), it was created internally by the Board, only by a few people.

And this "revision" has some really critical issues in it – like restricting the number of hummingbird feeders you can have, outlawing clotheslines (not very Green), dictating which flags you can fly, and restricting your choice of roof shingle colors to three (all dark). They wrap themselves in the magic cloak of "protecting property values", but honestly, is having three hummingbird feeders instead of two really going to damage property values? Do the many light tan to medium brown roofs in the community "damage property values"? Does drying my quilt in the sun destroy property values? They also added a section about how they can levy "Special Assessments" – closing the barn door after the horse is out.

The final insult is that the Board offers only a "yes or no" vote on the revisions, and voting now has several methods including online and by mail, which are not allowed in the current bylaws. That seems like lawsuit fodder. The "yes or no" vote option forces residents to approve bad stuff along with the good – in this case it is about 50/50. Sound familiar? Congress does this every year by waiting until the last minute, then demanding passage of an "Omnibus" package. This is how bad legislation gets through, that could never survive a single-issue vote.

The solution is for the residents to vote NO on the changes, in person, at the October 6 meeting (less chance of vote counting shenanigans with an in-person vote, since it is the HOA Board that counts the votes). If this hot mess of revisions is voted down, the incoming Board (with several new members) could open the process back up. They could send out a questionnaire, and put together a wider group of residents with a range of opinion, instead of just the "more power to the HOA" crowd. Then, a new, "clean sheet of paper" slate of rules, not the 50-year old mess, could be crafted. The community could then vote ITEM BY ITEM on the areas where the HOA would be granted power. This is a more open, fair, and current method, which the outgoing Board seems to loathe. Opening up the revision process gives the whole community a voice, not just a few with an agenda. During and after the new set is crafted by the community, it would be wise for an attorney to give the document a hard look.

It will be interesting to see if the Bolsheviks prevail on October 6, or if enough Mildweed residents care to make OUR rules, and the rulemaking process, more fair. Meantime, feed those hummingbirds all you can, in case the HOA wins and residents have to starve them out. Why does this HOA Board hate birds so much?

Even though this hypothetical situation is specific to a set of issues, the message is that if you live in an HOA, keep a hawk eye on their doings (if they allow hawks). Just like Congress, HOA boards seem to forget that they exist to serve their WHOLE community, not just the petty vendettas of individual HOA Board members.

EPILOGUE

At the raucous (hypothetical) Oct 6 meeting, the ridiculous changes to the rules were VOTED DOWN by the majority of residents! The pool boondoggle was called out, although the only thing accomplished was embarrassing the inept Board for their shenanigans. And the incoming Board got a message loud and clear that they would be under a microscope going forward. No more Karening, bullying, and opaqueness – the spotlight is ON. Ah, I do like it when a plan comes together!

The Science of Small-town Citizenship

I recently read an editorial by Kay Smythe in the Daily Caller, about how Finland may have the solution to America's mental health crisis. She made a great point that a lot of America's societal woes are because every child is told they are "special", which is impossible – the rule of the bell curve demands that the vast majority of people are "average". The messaging that everyone is "special" leads to an entitlement mentality, ultimately setting people up for disappointment and anger.

Finland is regularly ranked as the happiest country in the world. Sorry, Americans – our sense of superiority in everything is "proven" false by polls like this. But Finland - grey, cold, rainy, over-taxed Finland, seems to be onto something about societal happiness and resulting mental health. They employ three tenets in their society that we would be well advised to emulate. These three magic societal rules are – 1) A strong sense of community and "relatedness", 2) Doing good deeds for others, and 3) Finding a purpose in your life.

Let those three tenets sink in. The first thing I noticed was that they are mostly altruistic – focused on OTHERS rather than SELF. This runs in direct opposition to the "me, me, me" message of so much media (I spent years undoing the damage of programs like "Hannah Montana" on my offspring). The Fins seem to have unlocked a very simple magic bullet to happiness

– focus on others rather than yourself. Of course, we all must engage in some level of self care, but instead of 95% of your efforts in that area, consider halving that number, and spending some time thinking about and helping others.

So, how does this "love letter from Finland" translate to us here in Arizona's Mountain Mayberry? How can we translate this new outward, altruistic focus to the betterment of our small town? There are several ways that leap to mind. Serving in a philanthropic organization – Kiwanis, Rotary, Lions, Boys and Girls Clubs, Big Brothers / Big Sisters – and many more who offer direct assistance to specific needy groups in our community. Churches almost all have some kind of charitable outreach – I have been directly exposed to the good works of Catholic Charities as an example.

Involvement in governmental committees and boards is another great way to get involved to make our City and County better. Planning and Zoning if you are concerned about growth. Parks and Recreation boards are always looking for volunteer participants. Getting on the School Boards, or attending focused meetings about how our children are (or aren't) being educated, is critical for the future of our community. Attending public hearings and meetings about infrastructure projects is a great way to have your concerns registered, and considered by design teams. Sending comments to the transportation authorities about our too-low speed limits, or ridiculous roundabouts, can be a great way to register your input. Too many citizens spend their efforts griping on social media, which is kind of like shouting at the sky. I suppose people feel better by venting, but it is wasted wind since the decision-makers about the issue at hand generally don't hear it. Inasmuch as social media is convenient, and at our fingertips 24/7, it really isn't connected to the formal comment protocol.

Of course, no discussion of being a better small-town citizen is complete without mentioning the simple stuff. Random acts of anonymous kindness go a long way. A smile is a great way to start off any interaction. Simple manners (all too often receding these days) like saying please, thank you, yes sir, no ma'am, a proper handshake, eye contact, asking for and remembering someone's name - all contribute to the healthy fabric of small-town citizenship. I know, it sounds corny and obvious, but too often people skip the simple niceties of human interaction. Then there is the occasional opportunity to shine – for example I recently assisted an elderly gentleman get from his car to the front door on the icy parking lot of the hardware store. It took me maybe a minute, but his smile and "thank you young fellow" left me aglow for the rest of the day. I also didn't mind being called "young fellow" at the ripe age of 64.

I often hear that Prescott is "becoming Phoenix", people drive too fast, there are too many people, blah blah. I would counter that if Mayberry is small-town heaven, and Phoenix represents big-city hell, Prescott is a LOT closer to heaven than hell. We can keep that heavenly proximity by employing the ideals of the Fins, and by teaching and practicing good common manners. And side note, Phoenix is a poor example of "big city hell". I would offer San Francisco, Chicago, New York, or Portland as far better (worse) examples.

Also try to remember that some topics are "hot", and people get emotional or angry. Those interactions (frequent at public hearings about certain municipal projects) often turn ugly, where people tend to demonize an opposing opinion. The wise move is to take a breath, avoid vitriol, and stick to the facts and logic of your argument. Screaming and tears don't convince anyone – sober discussion of salient points is how understanding is born. And you may never agree with the opponent – but that's okay. The majority rules the day.

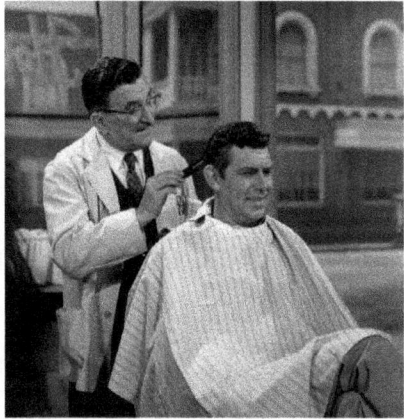

So, go out there and smile! Shake a hand, make eye contact, and use the other person's name. Do a good deed. Join a philanthropic organization or City Board. Choose your personal purpose. Start small if you have to, baby steps count, too!

167

The Science of Feeding Ourselves

One of the unexpected joys of living in our Mayberry-in-the-Pines is the Prescott Farmers Market. When we moved here, the PFM was located at a parking lot at Yavapai College, and that is where we first found it. For quite a while we patronized that location, mainly picking up a few garden goodies, a yummy baked item, some stellar tamales, or possibly just coffee. Our trips to PFM were occasional, and a nice Saturday diversion.

Then the rona insanity kicked in, the PFM was a place to get fresh food without being harassed at the chain grocery stores. We really ramped up our visits to PFM then, and continue to this day. We have always appreciated the idea of locally-grown food, and supporting local farmers and artisans. Kathleen Yetman, the Executive Director, should get some kind of award for running this fabulous service for Prescottonians and visitors alike.

In the ensuing years, we made an effort to visit all the booths, and our eyes were opened to the variety and depth of offerings by the vendors. My wife started participating in the composting service (we have heard it called the "Compost Church", a name which I rather like!) I have become a regular at the bread guy, located along the north row. He has no sign on his booth or truck – he doesn't need one. There's always a line. His Jalapeno Cheese bread is an addiction of mine. I don't even know his name – our bond is an unspoken one between Jalapeno Cheese bread pusher and addict.

I always look forward to fall, when one of the vendors (Whipstone Farms) cranks up the green chile roaster. The intoxicating aroma of roasted chiles spreads across the grounds. They are strategically located in the southwest corner, so that prevailing winds carry the aroma right over the whole Market, clever guy. I have been known to stand right in the path of that fabulous smell, and close my eyes. The scent transports me back to the 90s when we lived in Albuquerque, and our kids were little.

Grace and Rob run a stand called "Arizona Desert Rain". They make various salves, lotions, and CBD products, using a chaparral bush base. I swear, when I rub that salve on my arthritic thumbs, it smells EXACTLY like the desert after a rain! And the stuff really does provide pain relief (for me, anyway). I used to chase down Grace and Rob at markets in Flagstaff, and even as far as Pine, and was thrilled when they chose to come over to the Prescott market. I always enjoy a stop-by-and-chat, even if I'm not buying anything that day.

Another of our favorites is a stand that sells mushrooms. Forgive me for not knowing the name, but the "mushroom lady" is also on the north row, and her wares are fantastic. If you enjoy partaking of the culinary delights of the mycological world, she has some delicious mushrooms that will really make a dish. Our favorites are the Oyster mushrooms and Lion's Mane.

169

Whether as an addition to some pasta, flavoring for a steak or fish dish, or just sauteed and eaten straight, these fun fungi will grow on you.

Several vendors carry vegetables and fruits, nuts, honey, and lots of other staples. One gentleman has farm-grown meats and eggs. Another does an amazing job at knife sharpening. I bought an absolutely gorgeous cutting board for Sweetie for her birthday, made locally. I wish I had room to mention them all. But rest assured, you can get most of what you need there, and keep that money LOCAL!!

Something that the PFM also sells, but there's really no charge, is a wonderful sense of community. On any given market day, I stroll the grounds, usually running into several friends and acquaintances. Sometimes I see Barney there with Thelma Lou. Aunt Bee picking out flowers for a dinner table arrangement. Folks from my neighborhood. It is really hard to just stop and grab a few items in a few minutes – the social aspect of the Market is a valuable byproduct that takes some time. I also enjoy seeing strangers enjoy a coffee, listening to live music, or taking little kids around to experience new things.

A great example of the community-knitting that goes on was evident at a recent market, where the weather turned out to be windy. Some of the shade structures were trying to take flight, and shoppers instantly rushed over to help hold down the tents until some rope could be employed. One vendor inadvertently released a plume of fliers in the wind, and the crowd did a workmanlike job of running them down so they wouldn't become litter. The instantaneous outpouring of help blew me away (see what I did there)?

So, if you haven't been to the Prescott Farmers Market, or you haven't been in a while, stop by. They have a web page that gives hours – winter hours are happening now (Saturdays 9:30 to 12:30). Spring is when the variety of offerings really expands, but there's still lots available in winter. I will be looking for you! Be sure to bring your Mayberry attitude, and keep an eye out for Ernest T. Bass!

The Science of "Buyer Beware"

We have all heard the term "Buyer Beware." It means that a buyer should do his/her due diligence (research) before making a purchase, and take personal responsibility for their decision. A recurring theme here in the Yavaplex involves residents and homeowners belly-aching about certain conditions on or near their property, which pre-existed before their home was built. Specifically, these issues involve airports, mines, roads, fire risk, and water. Every one of these issues existed before they bought their property. So, let's take these one by one.

The location of Prescott's airport (call sign PRC) is well known. The airport was originally built in the 1920s (roughly 100 years ago), well before any homes were built around the airport. As of 2021, PRC was the 18th busiest airport in the country (surprise!) in terms of daily flights. That number is high because of our local point of pride, Embry-Riddle Aeronautical University (ERAU), which trains many of our airline pilots. The majority of daily takeoffs and landings are small propeller planes operated by ERAU for training flights. PRC also has commercial jet service from United Airlines, plus a fair amount of general aviation traffic, both jet and propeller.

So, what's the problem? Noise. Noise from aircraft is expected near an airport. In the 1950s until today, developers built residences near the airport. I wonder if people who buy those houses are aware that there is an airport nearby? Had they done even the most basic research (looking UP when scouting the neighborhood), they should have known that airport noise comes with buying a house NEAR AN AIRPORT. Also, they should have known that airports in growing areas don't get less traffic over time, they get more. So, simple solution. If you don't like aircraft noise, don't buy a house near an airport.

Rather than accept the characteristics of where people buy property, some people wish to shut down allowed property uses on property that they don't own. If you aren't ready to accept that mines exist or have existed near your little slice of paradise in the pines, perhaps you should have considered buying property where mining is not permitted anywhere near you. Again, buyer beware – the most basic research will show any prospective buyer where old mines are located, or where mining is allowed.

Roads are often a point of contention. If you buy a house next to a busy two-lane road, any reasonable person might expect that the busy two-lane might become a busy four-lane road in the future. In my work on highway projects, it always surprises me when people are upset by traffic noise, in a house they bought one street over from a freeway. Against my advice, family members have bought property next to busy roads, and then regretted it. They ask me (after the fact), what they can do about it. My answer is always the same – move to where the traffic isn't, if you can't adjust to it.

Forest fires are a risk issue if you choose to buy in a forested area. We all love the cool pines, the lovely wildlife, and the solitude that living in or near the forest provides. We also must accept ("Buyer Beware") that the forest may burn someday – it is a natural occurrence (and arson is an unfortunate non-natural occurrence). The best you can do to mitigate this risk is to "Fire-wise" your property, making it less likely to burn. But in reality, a big fire might take your house anyway, even with Fire-wise measures. This is a risk you choose when you buy in the forest.

Finally, water. For years, the local extremist activist group has been crying that a few private wells have gone dry in outlying areas, therefore we must all be terrified that we will run out too. Poppycock. When you buy a rural property, you should consider the aquifer, where your well sits upon it, and have a backup plan. If you are on the fringes of an aquifer, or the well on the property you are considering has a history of low output, you should expect water supply problems during a prolonged drought. Also, if you are out in the hinterlands on a single well, obviously you do not have the water security of multiple sources (you have chosen that risk). Government does not guarantee groundwater. Urban dwellers don't have this worry, because the city has wells in several locations, minimizing the single-point withdrawal risk. The Arizona Department of Water Resources

(ADWR) does an excellent job of providing groundwater information, and assuring a long-term supply. Again, "Buyer Beware," and if you choose to rely on one well, do your homework. And have a backup plan for when extended droughts (normal in Arizona) affect marginal wells.

So, in all of these situations, "Buyer Beware." When considering purchase of a property, look up. If you see and hear airplanes, it is a safe assumption that they will continue to make noise. Look at maps of where old mines are located, and where mining is permitted today. Look at roadway plans for your area, and observe traffic patterns yourself. If you stand in the back yard of a house you are considering, and you hear the roar of traffic, it will likely continue, and probably increase. And finally, if you are purchasing a farm or ranch or rural property, do some serious research about the well onsite, including ADWR information about wells in the area you are considering.

It is up to YOU to be informed, choose your risks carefully, and not be bitter and blaming after-the-fact. It's called personal responsibility. Try it, you'll like it! And you'll likely be happier.

The Science of Driving in Mayberry

It is easy to criticize other drivers, and to think that our own personal driving skillz are easily in the top 2% of all drivers. Given the immutable nature of statistics, we must accept that driving skill is measured on a bell curve. At one end (a minority), where we all would place ourselves, are the Mario Andrettis of the driving world – always attentive to the world around us, lightning-fast reflexes, and years of experience that make our driving prowess stand out among the other humans as the example to follow. The other end (another minority) of the bell curve consists of people whose sight, hearing, and reflexes peaked during the Carter administration, or young ones without experience or judgment, with their noses buried in their phones while they drive. The majority of the humans fall somewhere in between those extremes, in the meaty part of the bell curve. But, guaranteed, they self assess as being more like Mario than, say, Mr. Magoo.

As examples of driving skill, I will use my parents, whose driving styles were as different as daylight and dark. My Dad, a salesman in the oilfield for much of his career, drove around 90,000 miles per year (this is not an exaggeration). He learned that on those long, flat, empty highways of west Texas, he could get more sales calls in, and make more money, if he got from place to place quickly. Texas had no points system then, so tickets were simply a cost of doing business - and my Dad knew most of the judges in the area. His mantra, which I have adopted, is that a ticket is just your opportunity to pay a voluntary tax, and to thank a law enforcement officer. He was always focused, never distracted, and had the reflexes of a man who had been a three-sport athlete. Crashes? In the (literally) millions of miles that he drove, he had one wreck – at 3 AM coming home one night in 1959, he fell asleep at the wheel and hit a parked diaper service truck. No serious injuries, just a broken nose and two cracked ribs, two totaled vehicles, and fresh diapers strewn down the block. I always felt utterly safe with my Dad behind the wheel, and it is his driving style that I have modeled.

My Mom was a completely different story. Speeds were usually 5 over the always-too-low "limit", which is normal. Hands at ten and two, alert and courteous. The very picture of the Driver's Education manual. Her driving was like her perfect Palmer method handwriting – efficient, elegant even, and always proper. She never got a single ticket in her entire life, not even a parking ticket. Only one accident – when the preacher's kid neighbor zoomed out of their driveway without looking, and smashed the side of Mom's new Oldsmobile. That was also the only time I ever heard my Mom utter a cuss word.

Another example of a driving "style" was my paternal grandmother. She lived in Houston, where driving is a survival skill. Tempting fate, she always drove 10 below the speed limit.

She used the center stripe as a centering guide for her car. Four-way stops were for the other guy. The rearview mirror dangled by a string, because she only used it when applying lipstick. Every corner of her black '73 Impala was crunched. The Houston police asked her several times to please not drive on the freeway, where she held a steady 45 mph, straddling two lanes. As is common with many older drivers, she was reluctant to surrender the keys when her driving became an obvious menace to man, beast, and the city of Houston. My uncle resorted to sabotaging the distributor, and so the Impala sat for a year or so before being sold. My grandmother got used to bumming rides to the store, church, and beauty shop from her friend Berniece, who was only marginally better behind the wheel (luckily Berniece avoided the freeway).

I would go visit my Mom every couple of months in her final years – she had moved into a nice retirement community with her own apartment, with a porch for tomato plants, and she could keep her little dog. My job was to take the old Cadillac and get it washed and gassed. One trip, when I had been away for longer, I noticed that the Cadillac was dusty, had a low tire, and was still full from the last time I had gassed it up. I asked Mom about it, and she sighed, and said "just go ahead and sell it. My reflexes and eyesight aren't what they used to be, and I can use the shuttle bus for errands". I was thunderstruck, remembering my grandmother's story, and the stories of many friends with similar stories of parents and grandparents. My Mom, God bless her, made it easy for me.

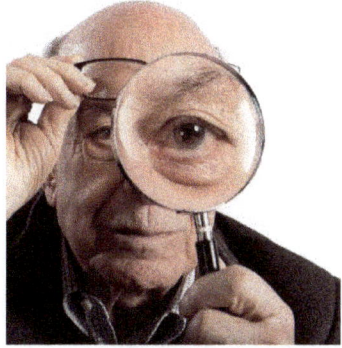

So, my stories have a corollary to living here in our Mayberry-in-the-Pines. We have a wide variety of drivers here, but more than our share of, shall I say, "well-seasoned" ones. Recently, on my short trip downtown, I had to dodge THREE ditherers. Two of them had no idea of the near-mayhem in their wake, and the third looked right at me, pulled out in my path, and smiled as I slammed the brakes and swerved.

Don't be the ditherer. Please self-assess, and ask a couple of brutally honest friends if they think you should keep driving. If you get honked at more than once a week, or you notice eight cars backed up behind you, there's your sign.

CHAPTER 6
Transportation Topics

INTRODUCTION

How we move about in our communities varies, depending on where you live. Most Americans use a private automobile. For longer trips, an airplane is the popular choice. If you live in a handful of big cities, commuter rail might be an option, inconvenient and dangerous as it might be (the other passengers, not so much the train itself).

Whatever your method of local locomotion, every type of transportation requires public land and facilities. In my career, I have worked on highways, local roads, freight rail facilities, mass transit rail, airports, and my favorite, ports (they are always hideously contaminated, meaning more work for me!)

This chapter lightly touches on these topics, and includes roadway congestion, the public involvement process, air travel, and the bane of my existence, roundabouts (also known as rotaries, traffic circles, or my favorite term Circles of Hell).

If enough of you read this book and I make it to a Volume 2, I will include LOTS more about transportation. I have a dozen or so transportation topics in the queue for next year's columns.

Congestion (cars, not noses)

Way back in my "Environmental Regulations" column, I laid out how well the Clean Air Act had cleared our air, even with many times more cars are on the road than when the CAA was passed about 50 years ago. Traffic congestion is common today in large cities and even in small towns. Although the Greater Yavaplex doesn't quite have the congestion problems of nearby Phoenix or other urban –plexes, I felt it would be worth a column to explore the issue, and offer some common-sense solutions and suggestions.

I worked for decades for companies that primarily did transportation engineering – highways, roads, airports, ports, rails. Working with these fine transportation engineers has not made me an expert in traffic engineering, but I have picked up some of their wisdom and madness over the years. For every marvelous, computer designed, perfectly engineered highway curve, engineers have an equal and opposite approach or term that is maddening. One particularly aggravating term is "storage." This is what they call a place where cars are

likely to slow or stop on an otherwise free-flowing lane. None of us like the idea that we are "stored" on a roadway, but alas, the engineer uses this term as a sterile description of a traffic jam, that makes the rest of us cranky and rage-able.

Surprisingly, the highway engineers' goal is not for every highway or city street to be an open, clear lane to speedway happiness. In fact, this condition (also called Level of Service A, or LOS A) is wonderful for the driver, but considered to be a bit of a waste of pavement to the highway engineer. They prefer a little denser packing of cars, to get the most travel bang for the highway construction buck. Of course, when we get to LOS C (Flagstaff on a weekend), or even the dreaded LOS F (Phoenix or Tucson at rush hour), things get a little less "efficient." Rest assured that for every time you are "stored" on a road, or when you are cruising toward Ash Fork at 75 enjoying the scenery, some anonymous highway engineer is studying ways to make your road function better.

The Post Office has a "Wanted" poster for folks that the local police would like to meet. I've yet to see anyone in the hoosgow for "road rage" (but it is only a matter of time). So, here are a few suggestions to help keep your face off the Post Office poster, and bring you some other benefits:

1. Think before you leave the house. That's right – use that lump of tissue between your ears a little bit. Ask yourself if the trip is really necessary, or if you could go

later or earlier. Timing of your trip is enormously important as a congestion factor. Maybe that trip to get bean dip and Fritos can wait until after the 5 PM "rush hour." Explore shifting your workday. If your boss will allow it, consider working from 6 AM until 3 PM, or 10 AM until 7 PM. This little change works wonders in blunting the peaks of "rush hour." Of course, "telecommuting," or working from home by computer, is the ultimate congestion-killer, if you have a job that allows that. Even a day or two a week makes a huge difference on the roads, on your sanity, and on your wallet.

2. Consider "chaining" your trips. If you need to go to the grocery store, the dry cleaner, the feed store, and the hardware store, do it in one chained loop rather than individual trips. I am reaching that age where sometimes I go into a room and forget why I went there (and this has happened at the hardware store too). Making a list is not a crime – make your trips planned and purposeful. A loop trip limits congestion, because your car isn't on the road as long, or as many times in a day.

3. Consider an alternate route. If everybody crowds onto the main road to go across town, that road will, by definition, get congested. But if you can go another way, even if it is a bit further but not congested, you've

done two good deeds – reduced congestion on the main road, and used unused capacity on the side street. Imagine if you had the option to get off SR 69 and take Sun Dog Connector!

4. Finally, the suggestion that never plays well in a small town - think about an alternate method of transportation. In big cities there are trains, buses, even rick-shaws. Every one of these people-concentrators removes cars from the road. In smaller towns the choices are more limited, but maybe a motorcycle, bike, or fast tractor will do. And remember, we are in the Rodeo Capital, so the good old reliable horse is a great way to help congestion while preserving our heritage. The horse will probably enjoy the change of scenery, too.

The net effects of these congestion-reducing ideas are many and wonderful. Less congested roads. Better air quality. Safer streets. Fewer road ragers. Saved gas money. Happier horses. But while all these effects are wonderful, the intent is not to relieve you of your beloved car. Some socialists want to take away our zoomy cars, and MAKE us ride a bus. Sorry comrade, we are Americans, and it is our God-given right to drive our V-8 chariots whenever and wherever we please.

But we can certainly make the shared road experience better by making smart choices, and keeping those roads less congested.

The Creeping Socialism of Roundabouts

Although I am a Geologist by trade, I worked for most of my career for a large civil engineering firm, and supported hundreds of roadway/rail/airport/port projects as part of the environmental clearance team. Through these projects, I learned about the thought processes of civil engineers (scary, right)? I have also driven approximately 730,000 miles in my life, so my qualifications for this article are unassailable (LOL).

We have all noticed the proliferation of roundabouts on our roads, for about the last 20 years here in Arizona. Traffic engineers speak glowingly about improved traffic flow, and the magic wand of "improved safety". I hate roundabouts with the white-hot passion of a thousand suns. I will explain why.

First, understand that roundabouts work best where multiple (3 or more) roads come together. Complex angles at intersections are never good, and we Americans are programmed to the idea of two roads intersecting at right angles, with a stop sign or a traffic light. In Europe, particularly jolly old England, roundabouts make sense. They don't have large gridded streets with well-timed lights. When you have thousands of tiny cars a day going 6 directions into one point, the whirl of a roundabout sort of works. For three good reasons – the drivers there have used them for their entire lives, the roundabouts usually have a large radius (which maintains speed), and they are well-designed with good signage. Here in

186

Arizona, those three criteria do not exist. Roundabouts are new (less than 20 years old here), and usually poorly designed, with a small radius. A perfect example is the seventyleven roundabouts on SR 179 between I-17 and Sedona through Oak Creek Village. WAY too small a radius – trucks and buses struggle to negotiate these tiny circles, slowing traffic. I happen to know that the radius issue was intentional, but that is an "over-a-beer" story.

The most ridiculous roundabouts are those with little to no side traffic (between Camp Verde and Cottonwood, for example). The only spokes on those wheels are SR 260 each way, and maybe an unpaved cow trail with its own ramp. Locally, the abomination that was recently completed on Pioneer Parkway in Prescott fits this model. There is very little side traffic, not even enough for a traffic light before. And no voluminous data set of horrific crashes either – everybody pretty much had the old Pioneer Parkway / Commerce Drive intersection figured out. It didn't even warrant a traffic light, and it worked great.

So, why these roundabouts with no side traffic? I have had traffic engineers tell me, in a moment of unguarded candor, that the roundabouts are there "just to slow us down". Because slow equals safe, right? WRONG. Any traffic engineer worth his/her salt will tell you that speed DOES NOT kill, DIFFERENTIAL in speed does. If the "slow is safe" trope were true, then the state should just implement a statewide 20 MPH speed limit, everywhere. Anyone who has driven in Prescott knows about speed differential being unsafe – the two most common speeds driven simultaneously on the same road are 25 and 52 (I'm looking at YOU, Willow Creek Road). Dangerous.

Another problem is ridiculously low speed limits, and again I cite Pioneer Parkway. It has a design speed (speed that the road is designed to handle) of 65-70, but the posted limit is 45 in places. Lacking speed limit signs, a normal driver will drive a reasonable speed, mostly governed by the design speed of the road. On Pioneer Parkway, a reasonable speed is about 60, which is how fast about half the traffic drives. So, why the artificially low limit of 45? The County says it is because it is a "residential area". Really? Then why is an Interstate-worthy corridor going through it, and no driveways face the road? The new 20 MPH maximum roundabout in the middle of this previously well-flowing arterial will surely generate speed-differential mayhem.

Two other things about roundabouts that engineers don't like to talk about is noise and pollution. Every car slowing down, and then speeding up, creates more noise than a car going a constant speed. It also emits more pollution with the slow down/speed up cycle. Yes, traffic lights still have this effect, but not on EVERY CAR. With a properly timed traffic light, most traffic has a good chance of hitting the light green, and never slowing down.

The socialist aspect of roundabouts comes from the way they steal time from EVERY DRIVER going through it. Total up the time lost by every car having to slow down, and the lost productivity becomes substantial. With a well-timed light, you have a good chance of not slowing down at all. But traffic engineers are not measured by how aggravating their designs are, or how much time they steal from everyone. Equalized suffering is socialism.

Then there's the old standby magic wand, "safety". Engineers tout roundabouts for "trading low-speed bent metal for high speed T-bone crashes with injuries" (although the data doesn't show much difference). Tell that to the truck driver who flipped on the roundabout because his load shifted. Watch the video of roundabouts when it rains or snows. Engineers don't talk about this. And how many times has someone fouled your lane in a roundabout? Roundabouts require driver discipline and skill, which is a rare commodity.

Dogs, cows, and horses also HATE roundabouts (I'm in good company). My dog gets thrown left/right/left through every hated roundabout. Cows and horses have to do it standing up in a trailer. I would love to put an unbelted traffic

189

engineer in a horse trailer and drive from I-17 to Sedona. He would be unconscious by Bell Rock.

One final thought – if roundabouts are there for "safety", then use another safety feature on your car to double up the safety. Blow your horn all the way through every roundabout, every time. You know, for "safety". The neighbors will love that roundabout even more.

Rocks, Roads, and Connectors

Lots of discussion is going on these days about roads. Last week saw a public meeting about the Sundog Connector, which will provide an alternative connection between the two largest towns in our county. This week we will have a public meeting about widening of US 89 through the Dells in Prescott.

One meeting has, and one meeting will, provide the public with an opportunity to weigh in on these transportation projects (although the public

U.S. 89 through Granite Dells - Prescott, Arizona

doesn't actually get a direct "vote", a common misunderstanding of the process). In my career working for a large transportation engineering firm, I participated in dozens of these meetings. They can be informative, contentious, enlightening, and frustrating, all at the same time. Some people go to ball games for entertainment, I like to go to public meetings, because they are a great way to observe human behavior - good and bad. I am also a big fan of nature, National Parks, and especially pretty rocks. But I also know that we humans are part of nature, too, and some kind of balance must be considered. Read the preamble to the National Environmental Policy Act (NEPA). The

191

law that gave us 7000-page Environmental Impact Statements (EIS) mentions "balance" numerous times.

It is important to understand the players at these public meetings. In charge of the meeting is the state agency, city, or other governmental entity that owns the project. The meeting is usually put together by their staff, or a consulting firm hired to do preliminary studies. Also present are elected officials, who make hay on being for or against such projects, depending on how the political winds blow. Local residents also show up, some who have a direct interest in the project due to proximity. Activist groups also show up, with their various axes to grind. Always entertaining are the NIMBYs (Not In My Back Yard), who might see the benefit of improved transportation, as long as it doesn't affect them directly. The most fun are the BANANAs (Build Absolutely Nothing Anywhere Near Anything). These folks basically hate humans (especially new residents), but like living here.

We have seen all of these players weigh in on many recent projects in the Greater Yavaplex. The proposed hotel on Whiskey Row. The "airport overlay", which is an attempt for the city to protect our vital air link from being crippled by encroachment on runway clear zones. Various meetings about water, because as westerners, fighting about water issues (real or imagined) is in our DNA. School curricula, zoning fights, you name it – we have our fair share of vitriol, appropriate for a growing area with passionate citizens. These processes are healthy and necessary, if a bit frazzling at times.

As for the Sundog Connector, this project seems like a no-brainer. With only one direct route between our county's two largest cities (Prescott and Prescott Valley), we have all seen the growing problem with congestion on SR 69. Emergency response times are impacted, bad crashes occur, and daily aggravation is something we hoped to avoid in moving to our twin Mayberrys. Sundog would help address connectivity between our towns, provide a safety valve if one of the two gets clogged by a crash, and would improve response and travel times. But the opponents claim to be upset about the loss of some grassland, and the usual wringing of hands about growth, with the attendant fear and loathing of "the evil outsiders". The proposed route traverses grassland that isn't particularly unique or scenic, and doesn't harbor any specific habitat for endangered critters. And I wonder, did the people concerned about this new road protest about the land their house was built on? Or the land that the Courthouse Plaza was built on? Or the land our schools and colleges were built on? Civilization requires space, and our towns are inexorably growing into one Yavaplex, like it or not. All of the protestors about growth are part of it. The hypocrisy is stunning.

The US 89 widening project has brought out naturalists, clutching their pearls about the "destruction of the Dells". The meeting this week will provide maps showing the proposed areas for widening. I have done some measuring on my primitive GIS system, and from my rough calculations, about 3 of the Dells ~3300 acres will be disturbed to widen US 89 at its 1.5 mile pinch point (the consultant on this project can check my math). That's about 0.1% of the Dells total area. And the strips of rock removal are located along US 89, an already disturbed area (do people really recreate within 25 feet of US 89?) As a geologist, I can assure you that there are more beautiful rocks right behind the 20-foot strip that will be removed. And part of the discussion is moving some of the boulders (that can be salvaged whole) to the end of the outcrops, extending the rocky beauty.

The thing to remember about both of these projects is that public safety is the main goal. The alternative to building Sundog to relieve SR 69, and widening US 89, is to leave things as they are. Traffic and crashes would continue to increase. The BANANAs don't want to see any changes at all. Well, change is coming. And stopping these projects will only result in inconvenience, crowding, and more mayhem and death on our roads.

I don't know about you, but I would rather rearrange some rocks, or replace a small stretch of grassland with a road, than go to my neighbor's funeral, or that of his kid, who will die in a crash in 2029 on a road that wasn't improved. At least the people stopped in traffic while the first responders remove the bodies will get a nice view of the rocks while they wait.

UPDATE

Gutless politicians have killed the Sundog Connector project for now, caving to a vocal minority and allowing the rest of us to be endangered on the overloaded SR 69 for more years. The US 89 project is now only considering spot improvements (turn lanes) to the two-lane configuration, and is not considering a four lane option. This means that any option chosen will be like putting a Band-Aid on a chain-saw wound. The lesson here is that people should vote for politicians with courage, who will make decisions that protect the MAJORITY of travelers.

The Science of Air Travel Today

Forgive me if I occasionally sound like I am from an older generation in this column – I am. But I recently took an air trip which really made me re-evaluate how to move about on this planet. The missus and I took a long weekend trip to rural Iowa, to celebrate my mother-in-law's 85th birthday. I am lucky to have a mother-in-law for which all the "evil mother-in-law" stereotypes don't fit – she is wonderful, a joy to be around, and an inspiration. But getting to and from Iowa was an unpleasant experience.

First, let me establish my travel bonafides. I have visited all 50 states and several other countries. I have driven hundreds of thousands of miles (give or take 730,000). When I worked for "the big company", I did 60+ flight segments a year. I started flying in about 1976 as a teen, first on Braniff, then as a dedicated Southwest Airlines flier. My peak travel year was 2014, when I flew 78 segments on Southwest. I used to enjoy flying – the convenience, the friendliness, the reasonable cost, and the wonder of getting in a metal tube and time-traveling to Bozeman, Miami, or Indianapolis in just a few hours. Not anymore.

With the deregulation of the airline industry in the late 1970s, flying became democratized, and air travel was no longer just for the wealthy. Airlines like Southwest, PSA, and

People Express sprang up, offering really inexpensive airfares. When I was in college, I could fly from Denver home to Midland for scarcely more than the cost of gas. And the planes were fun, with a camaraderie of people having an adventure, and flight crews who made it fun (at least on Southwest).

My recent trip included a drive to PHX to catch a Southwest flight to Minneapolis, then a drive to northern Iowa for the birthday celebration. It was a direct flight there, then coming back we had to stop in Denver, one of my least favorite airports. But I had Southwest points to use up, so we bit the bullet.

Problem #1 – filthy, crowded, expensive major airports. PHX Sky Harbor used to be a gleaming, newish, not crowded, clean airport. No more. Its condition on a summer day was about equal to a 1969 Greyhound bus station after a protest march. Barely a seat to be found while waiting to board, filthy restrooms, and the capper was the cost of food. A plastic-wrapped tuna sandwich triangle was available for $14.95. Now, I can afford food, but I'm not Elon Musk here. I ended up settling for an egg and cheese breakfast burrito for $10. And my wife's coffee was $9. Are you kidding me?

Problem #2 was my fellow travelers. I got walked into THREE TIMES by people walking while staring at their phones. Then, they got on the plane, and blocked the aisle once they got to their row, so they could fiddle with their bag and get out some Cheetos, while holding up everyone else trying to board. Then, my late-boarding center-seat seatmate was - how can I put this delicately – "generous of girth"? I spent the three hours to Minneapolis tilted 15 degrees toward the aisle (this did my back no favors), to accommodate the flab that flopped over the armrest (of course, no apology). Upon arrival, the deplaning experience was like people had never gotten off a plane before. Messing with bags, trying to go back against the flow to get a bag in the back – bedlam.

Problem #3, also a fellow traveler behavioral problem, had to do with the window shades. 80% or more of people in window seats close the shade for the whole flight. First, it is nauseating to not be able to get your bearings on a horizon. Second, why would anyone choose to stare at a screen, to watch a Friends rerun that they've seen 14 times, over a bird's eye view of God's creation? An amazing view is available most of the time from an airplane seat, yet screen-addicted zombies today can't be bothered to look outside.

Am I asking too much here? Watch where you're going. Board and deplane with efficiency, to get your fellow traveler on their way without delays. Open your shade so others can see out. Clean up after yourself. Did the modern traveler skip kindergarten, where most people learned basic courtesies? Are parents today so absent that they don't teach their kids (eventual grownups) how to act in public?

I hadn't flown in almost three years, because of the forced rona muzzling. I did a lot of long road trips. I think I will go back to that. I will also choose United when I do fly, from our clean, convenient local Prescott airport. The downside there is that I will have to connect through one of Hell's waiting rooms, LAX or DEN. I am also letting relatives know that pop-up trips for funerals and weddings are out for me, unless I have a week's notice to drive it. And anything east of Texas or north of Salt Lake City is out. Of course, for international travel, I will have to gird my loins and deal with a DEN or LAX connection. I guess paying for First Class, and drinking a lot, will make that tolerable. But only for big vacations.

Okay, rant off. I am really disappointed in some of you people, and your parents need to be alerted to your travel failures. For those who say "excuse me", make an effort to hurry along, and fit in a standard airplane seat, I thank you.

CHAPTER 7
Fun Topics

INTRODUCTION

This chapter is the potpourri, the hotdish, the buffet of topics that don't fit anywhere else. Writing a science-based column, and reading it if you are a subscriber, would get boring fast if the topics were always heavy science stuff. So, I try to sprinkle in fun topics, things that I just want to write about – and my editor indulges me on this.

So, in this chapter I have a few that were written for publishing just ahead of a holiday (Easter, 4th of July, Thanksgiving, Christmas). The buffet also includes a column about social interaction with the other humans, "prepping" for when the other humans cause big problems, getting away from the humans to beautiful, isolated places, and a couple of columns just for fun.

If I have learned nothing else from my life of science, consulting, logic, and common sense, it is this - the moments of mirth, the silly stuff, the fun with friends, the antics of my dogs, the giggle-til-your-sides-hurt moments are the soft chewy center of this confection called life.

Science and Faith

In the spirit of Easter, I am laying aside expensive eggs for the spiritual side of science. What, you say? Aren't science and faith mutually exclusive? Au contraire. I will make my case for a robust discussion of science and faith, knowing full well that I am flouting the conventional wisdom.

In this column, I often take a current topic and try my best to provide basic information on scientific topics, free from spin, using the "Scientific Method". For the uninitiated, the Method involves forming a hypothesis, and performing analysis or experiments, and presenting "evidence" to prove or disprove that theory. Lawyers do this in their own way, building cases using comparisons to existing case law (basically legally-hardened opinion) and observable facts. Scientists use facts, empirical truths that can be measured and shown, and most importantly, repeated with the same result. We scientists like math, because numbers are not very flexible (unless money is involved). We like chemistry, because it also is a science that clings hard and fast to a reproducible result. Although it isn't hard to lean one way or another when building a case (by selecting methods that might favor your result, also known as Activist Science), the Actual Scientist has to "show his work" and prove the position.

This is why I always look forward to Easter (and Christmas). These are times when science goes on hold for a while, and we operate on faith. Faith is a funny thing – it lives somewhere between science and wishes. Faith is as powerful an argument as any for a given position, in that people believe their faith-indicator more strongly than any argument a scientist (or a lawyer) can provide. As a scientist, my "comfort zone" is to try to explain the unexplainable event in terms of science, but in many cases, I am left stammering. Take, for example, the power of prayer. The scientist in me says that voicing or thinking a wish, in the form of a prayer, is a singular act, contained within the cranium of the person doing the praying. How can that possibly have tangible effects beyond the feelings of that person?

Yet I have personally experienced the positive effects of the power of prayer. Unexplainable outcomes. People with illnesses that should be fatal, yet against the diagnoses of multiple doctors, they recover. Events that could not even be wildly imagined, come true. But prayers can run both ways – sometimes that which is most fervently prayed for, is denied. In many cases, the unanswered prayer turns out to be the best outcome in the long run. Who is in charge here? To the scientist, this is an unpredictable system. But it somehow seems to work just fine. That is the thing about faith that sometimes makes the scientist crazy – *how can we know (or prove, or predict) the outcome?*

The answer is that we can't know all the answers. As a geologist, I am lead-pipe certain that our Earth is around 4.5 billion years old. Sand is deposited at a certain rate – volcanoes flow at a certain rate – one trip to the Grand Canyon should prove to anyone that the Earth has been around for billions of years. Yet some people insist that the Earth is only 7000 years old, and that man walked the Earth at the same time as the dinosaurs. I can talk myself blue proving my side of the case, but the 7000-year-theory believer cannot be swayed – their language of decision-making is in terms of faith, not scientific facts. So, I don't get into those conversations – I just try to find the common ground that we can all share. "Isn't the canyon beautiful in this light?"

Everyone, scientist or not, can agree that sunsets can be indescribably beautiful. A forest can be cool and green and dark and lovely. The stripes on a tiger are mesmerizing, colorful, and can signal impending death. The sound of a babbling brook is a salve for the soul. Waves at the beach can wash away your cares, even if temporarily. And most importantly, **this cannot all be an accident.** Some form of higher being must be involved, because the beauty and mystery and splendor of nature cannot be a random laboratory accident. It is immaterial whether the hand of this higher being is involved at the daily level (hmmm, this sunset needs a little more orange – okay, **poof** – perfect!), or has just set up this experimental lab we call Earth, and checks in occasionally (hmmm, dinosaurs? I don't think so. Here's an asteroid strike. Do over).

What we can all agree on is that the surface of this beautiful, wet, blue marble where we live is someone's great work of art. As a scientist (and an amateur photographer), I am awestruck on a daily basis by this place, and I look forward to meeting the artist someday. Easter and Christmas always remind me who the artist is, and the day gives us all something to think about and celebrate. We should think about this and celebrate it *every* day.

Science *or* faith? Why choose? A double helping of both for me, thanks! And spare me your hate mail this week – nobody is changing anyone's mind. Both my faith and my scientific method have taught me that.

The Science of Fireworks

Most of us love the splendor, excitement, and color of fireworks displays. I have fond memories as a teenager of sneaking out onto the golf course of my hometown (Midland, Texas), near the launch point for the city's 4th of July display, and laying on the 5th green, looking up at the explosions of color. I have always been a big fan of explosions – for years, my dream job was to be "High Commissioner of Blowing Stuff Up". But alas, it was not meant to be. This column will look a little deeper into the science, history, and future of fireworks. Feel free to amaze your friends with these details, Cliff Claven!

The Chinese discovered gunpowder around 200 AD, and by 600-900 AD they had found many uses for it, including decorative displays. Early fireworks were basically what we would today call "fire fountains" – gunpowder, sand, and lumps of iron clinker put into an inverted cone, and ignited. These sent "fountains" of sparks up to 30 feet into the sky, and amazed the people of that time, I am sure. Jumping ahead to around 1830, we arrive at the generally accepted birth of the "modern era" of fireworks. Shells were filled with gunpowder, lumps of metals, and fuses, and launched into the sky via mortar tubes before detonating. The elements of the modern fireworks shell include a lifting charge (propulsion only), a timed fuse (delaying the burst), and lumps of various burning metals to provide color. In the early days, colors were limited to yellow and orange, since iron clinker lumps were the primary colorant used.

Into the 1900s we go, and clever fireworks chemists began experimenting with other metals. They learned that Barium metal gives a green flame. Sodium gives a bright yellow. Phosphorous gives us the bright white/silvery shine. Strontium burns a bright, deep red. Various formulations of copper provide blue, aqua, and blue-green shades, and copper/cobalt mixtures give us deeper blues. The entire rainbow of colors can be achieved by mixing these and other metals into flammable balls in the explosive shell. An effect that I particularly like is a star that shoots out from the burst, and changes colors as it goes.

This is done by layering of the metals in the ball. Also a hoot are the ones that exit the boom in a crazy flight pattern, and make a screaming noise (done by having "fins" on the balls, and

having holes in the ball to create a whistling sound). My family's name for these are "screaming Mimis".

This was the way fireworks were for most of my life. But in the last 25 years or so, serious technological advances have further refined the fireworks game. Computers now control the launch and timing of bursts, to the point that they can synchronize bursts with music, and compensate for factors like wind. The way charges are constructed has also made for great advances. Company logos, words, or shapes can be created using timed fuses, and spaced arrays of color balls within the shell (some with secondary timed fuses).

More recently, technology has entered the field of fireworks, giving amazing effects. The Beijing Olympics dazzled with an airborne army of drones, outfitted as tiny launch pads, and computer controlled using GPS. This created a 3D effect of color and patterns in the sky, showcasing where fireworks shows are going in the future.

Even more innovation is coming in other areas. Fireworks are known to spook/scare animals, and also pose a problem for people with PTSD, autism, or a range of other issues. "Quiet Fireworks" are being developed which have a much less booming sound. Daytime fireworks are also a developing trend, with bright enough burns to be seen in broad daylight, or even better, against a cloudy sky. Another area of innovation is "color washing", an effect which creates diffused color "walls" rather than bright points of light.

Fourth of July displays were originally meant to mimic the "bombs bursting in air", a line from the Star Spangled Banner. I rather prefer the more traditional displays, perhaps with a few of the modern WOW effects sprinkled in. Other events are more appropriate for the fancy stuff. Call me a traditionalist, but I like the old-fashioned-style show for Fourth celebrations. Leave the technical wizardry to Vegas, Disneyland, or Olympics openings.

Fun fact, the largest consumer of fireworks in the world is Disney – not surprising given the extravagant shows put on nightly at Disney parks. I have a friend who grew up in Anaheim, close enough to Disneyland for his family to watch the nightly display. As an adult, he moved to another state, and was kind of disappointed that not every city has nightly fireworks shows. Personally, I prefer keeping them occasional, so they maintain their specialness.

So, enjoy fireworks displays, with greater knowledge and understanding! And be sure to keep your pets inside, with the drapes drawn, and a TV or radio going to distract them. Until all fireworks shows are "quiet", let's be responsible pet owners as well as fireworks enthusiasts!

The Science of Thanksgiving

By now, you have made it through all the cooking, the main meal event, the football games, the family, the tryptophan coma, and Black Friday (if you partake of that madness). Today you are getting a bit tired of the Turkey Soup Deluxe, the Turkey Tetrazini, and the Turkey Hotdish leftovers. You might be trying to push some Tupperware containers of this stuff on the relatives, as they pack up to leave. And of course, trying to get someone to take the cranberry relish is an annual struggle.

I wonder how many people stopped to reflect on the original meaning of the holiday. In 1621, the Pilgrims had a three-day feast celebrating the harvest, and invited the Wampanoag tribe, the local Native Americans. The feast was peaceful, as both the Pilgrims and the Native Americans broke bread to give thanks for a successful harvest, and the blessings of the year. The dishes were different from what we eat now, but the preparatory work, the ceremony of giving thanks, and the shared humanity of surviving another orbit of the sun are the same as we celebrate now. I can only imagine Mama Pilgrim trying to offload some leftovers to the Wampanoag visitors with some Tupperware as they departed.

Today, we seem to get caught up in the modern trappings and contrived mechanics of the holiday. Most families have attendees who travel serious distances to be together. Long drives, or God forbid, flights on the most crowded airport days of the year, are an unpleasant part of the event (much more stressful than the stroll through the forest that the Wampanoag experienced). There is the frenzy of cooking and food prep for a few days before the meal (and especially the morning of), which probably had the same vibe in 1621 as it has today. Instead of music and the smoking of pipes (1621), today the entertainment consists of football games on Thanksgiving afternoon.

Something new since 1621 is "Black Friday", an invented day where merchants supposedly go from being "in the red" for the year to "in the black". The reality is that it has become a free-for-all at the stores, with mankind often displaying the worst of their public behavior (Black Friday Black Eyes). You will not find me anywhere near a retail outlet on the day after Thanksgiving. I like to take that day to go to the range and shoot, take a hike or drive in the beautiful Bradshaw Mountains, or (more likely) enjoy a long walk with the dogs then a nice nap.

Then the weekend comes, and it usually includes taking the out-of-state relatives to some amazing Arizona vista (Grand Canyon is only two hours away, or Sedona, which has become a favorite). We also use this day to whittle down the leftovers, including the Costco pumpkin pie that is the size of a trash can lid. Next thing you know, it is Sunday, and the relatives are packing up for their day of hell at airports or highways. We regain the house, and usually collapse for (another) nice nap once things quiet down (I sense a theme. . .)

But how many of us stop amid the whirl of activity, and really think about what "giving thanks" means? I take time to reflect on the blessings in my life – my loved ones, my dogs, my tortoise, the ability to travel around our incredibly beautiful and clean country, and simple pleasures like Tabasco sauce. Yes, I have actually given thanks for Tabasco, and deservedly so. And who am I thanking? In my case, God, but different people have different ways to express gratitude. The important thing is that you do, and that it comes from the heart.

I also take time to think about those who struggle. There are many local service organizations in our town, who work hard to help support deserving charitable organizations in our community. Many of the people who are helped have a much shorter list of things to be thankful for, which amplifies their thanks for the blessings they can count. Those folks are always in my mind, whenever I give thanks for the multitude of blessings that I enjoy. And if you have the time and inclination, consider joining a local service organization. I have always considered service in one of these organizations as a form of "Thanksgiving".

In closing, I have a suggestion for you to try. Every day, regardless of what the calendar says, take a moment to give thanks for your blessings. I find that a moment of gratitude (for me, in the morning) makes for a better day, and helps to adjust my attitude toward the positive. We all need more positivity these days, with negativity bombarding us through the media (and sometimes our fellow humans) constantly.

And I wholeheartedly suggest dinner out on Saturday or Sunday night. Make it Mexican, or Italian, or Asian, or barbecue, or Indian food. Anything but turkey.

The Science of Christmas Magic

This has not been a typical year. Usually the Christmas spirit takes hold of me within a week of Thanksgiving. The decorating of the outside of the house is usually my gig, but recovery from major back surgery means that my better half has been responsible for the outdoor work, as well as the amazing job she does every year with the tree and all the trimmings indoors. My shopping this year somehow got done very early, so I don't have the Christmas cheer trigger of shopping for gifts. And I haven't been bombarded by Christmas music, since I am very careful about my radio channel choices on satellite radio (I have discovered which stations NEVER play Mariah Carey).

It's not that I didn't want to get into the spirit on the normal timetable. A heavier-than-normal consulting dance card, and the preparation of this book gave me lots of distractions. Realizing that I was slipping into Scrooge territory, a week ago I decided to do something that I knew my back could handle, something from my childhood that was always a treasured memory. I went for an evening drive, seeking out cheerful lighted Christmas displays on houses. When I was a kid in my hometown in Texas, it was a special treat for me and my sisters to get into our jammies, take a pillow and blanket, and load up in the back of my Mom's Oldsmobile. My hometown was about 65,000 then, and just big enough to offer lots of variety in displays. From the wealthy enclave of Racquet Club, where the residents would pay the power company 50 cents per bulb (!!!!) to illuminate their trees (all white), to the poorer neighborhoods who seemed to feature lots of wooden manger scenes, the full spectrum of Christmas displays were available and could easily be seen in an hour or so. A couple of jewelry stores downtown, plus Penneys, several oil companies, and a couple of the banks, did elaborate window displays – our little taste of big city Christmases.

Prescott offers a few neighborhoods who really go all out, but more often it was one or two per block. But that's okay. I still got the same warm fuzzy that I got as a seven-year-old. Of course, the display around the courthouse downtown, blows away anything my hometown offered, and Prescott really does downtown right. Acker Night, with the multiple musical acts and venues, and people walking the courthouse plaza with wassail, is always a "Bedford Falls" moment, especially in my best ugly sweater.

But the lights, the gifting, the hustle-bustle, and the endless Christmas music don't really do it for me, and never has. Of course, when my kids were little, I would climb up on the roof and stomp around, with a big string of jingle bells. I went all out on the house decorations (about a 2/3 Clark Griswold).

And gifts and traditions were many. Such are Christmases when your kids are little. But for me, the best part of Christmas was always today, the 24th, in the evening. Our family does a hybrid Christmas Eve/Christmas morning gift program, but the part of the evening I am talking about is after that. Once the gifting is done and the tamales are eaten, the rush is over, and quiet settles in. That is the best part.

The quiet of a manger, with a light snow falling. The reason for the season. On Christmas Eve I will enjoy a nice fire in the fireplace, some quiet, some traditional Christmas music down low (not a single second of that Mariah Carey abomination), and perhaps a little eggnog with my beloved. A quiet night of reflection, free from the noise, the bustle, the news, the conflicts, the chores, and the contrived cacophony of a commercialized Christmas.

This quiet island of peace and joy usually lasts through Christmas morning. Then the cooking begins. We let our kids choose our forever Christmas menu when they were 4 and 6. Shockingly, they took the task seriously, and came up with lobster, crab legs, steamers, and artichokes! No kidding, this was what they decided we would eat on Christmas Day, in the year 2000. My wife and I fully expected them to suggest pizza, hotdogs, and cookies. We were floored. But grateful – this is still a very good way to fill up on Christmas Day.

Then we usually have a spirited gin rummy tournament, or perhaps Scrabble. Nothing electronic on Christmas Day.

From about noon on the 24th, through the 25th, we enjoy an island of quiet and peace, to enjoy each other's company and reflect on the reason for Christmas, and our blessings. Because the 26th always comes, with the rush back to "normal life", travel, work, and the other distractions of daily living. A day and a half out of 365 doesn't seem like nearly enough.

I wish I could bottle the feeling of peace and contentment that envelops me on Christmas Eve and morning. Of course, if I could sip from that bottle in, say, April or August, the real thing in December would cease to be so special.

Merry Christmas to ALL, and may you experience the joy, peace, and magic that the season is all about!

The Science of Social Interaction

The complexities of societal interaction are well beyond my depth professionally, and I have no professional training in sociology. Let me get that out of the way right up front. But I have spent over six decades on this blue marble so far, and I have interacted with thousands of other humans, from a broad variety of cultures, countries, walks of life, viewpoints, colors, creeds, and football team alliances. I have learned through these experiences that face-to-face interactions are very different from telephone conversations, written exchanges, email exchanges, or especially social media exchanges. The best analogy I have come up with as we go through life, is that meeting the other humans should be approached with the same caution and attention as a walk through the wilderness. A zoo is an inaccurate characterization, since you are physically separated from the zoo animals, and they are not in their normal habitat. This column will attempt to use a walk through the wilderness as an analogy for humans with whom we meet and interact.

First, let me say that I generally approach the other humans as if they are a friend I haven't met yet. Perhaps that is my Texas upbringing coming through, or my generation, but I am ever hopeful that the animal I meet will be as friendly and open as I try to be. I have been bitterly disappointed on many occasions, but I choose to continue my optimistic approach. I am a firm believer that if you start out on a positive note, an interaction is likely to continue in that vein. In this wilderness example, I am a dog, meeting another dog, with wagging tail and curious excitement. Horses are often this way on first meeting, but can be skittish. Again, some people are like this, too.

The vast majority of people you interact with, or pass on the street without interacting, are generally benign, neither offering pleasantness or harm. These critters are like deer, or cows, or birds, or rabbits – generally indifferent and just going about their day. Some people you meet are quite unusual, like a giraffe or a raccoon. They are generally harmless, but can be mischievous (in the raccoon's case). Some people are goats – seemingly benign, sometimes friendly or playful, but just as likely to head-butt you just for the sport of it. I think goats really have no malicious intent (they aren't going to eat you), but they are just kind of dense, and go with their emotion of the moment. Surely we have all known people like that.

Some people just want to be left alone, and will move away unless provoked, after which they might strike. The rattlesnake is a perfect example. I have known many people who give off signs that they don't want to engage (a rattle), and if you push it, they bite. This analogy works except for some snakes who don't like you, and are aggressive about it. If you have ever been chased down a wash by a sidewinder, you know what I mean. Rhinos and hippos also fall into this category. Big and round, they seem kind of jolly. Until they charge at you. Not jolly.

Other people act like hyenas or wolves. Loud and obnoxious, but their furry look and quadra-ped stance might fool you into thinking they are just a funny-looking dog. The difference with hyenas and wolves is that they are generally malicious, and operate much more aggressively in a pack. If you have ever been ganged up on by a group, you know what the hyena mentality is like. Human riots have this vibe, as well as some activist groups.

Most dangerous are the predators – lions and tigers and bears, oh my. Many of us had cuddly stuffed toys of these animals as children, leading to a false assumption that furry equals friendly. Humans that follow the tactics of the predator will often ignore you, or quietly draw you in by looking cuddly, then attack once you are within their radius of lunchability. Creatures like this will eat you, and not think twice about it. I think we have all encountered bullies and toxic individuals who thrive on attacks. My friend's cat is, I think, a tiger under the surface. Quietly purring one minute, then suddenly shredding my leg in a frenzy of scratching. I wear long pants to their house now. Unfortunately, long pants don't help with this kind of human.

All of my experiences with two-legged critters have taught me some valuable lessons. I go in friendly, willing to give any human animal a chance. The actions and reactions of each human determine whether they will become a friend (an Irish Setter to my Labarador-retriever-like personality), someone to just acknowledge (giraffe or elk), or someone to studiously avoid (tiger, rattlesnake, or hyena). Of course, the worst one is the one whose threat is muted. Some people are spiders or mosquitoes, who sneak in and inflict their venom unannounced. Those are the worst, the ones who remove any chance for human interaction and negotiation.

And don't even get me started on scorpions, fire ants, man-o-war, or sting rays. I have tangled with all of these, both animal and human varieties, and all I know is that I never want to see them again.

As you can tell, I will never be Marlon Perkins.

The Science of Being Prepared

As a former Boy Scout, and a student of the movie "Red Dawn" (the original) I have always taken to heart the slogan "Always Be Prepared." "Failure to Prepare is Preparing for Failure" is another way to think about it. Many stories and articles have been written about the "Prepper" lifestyle, but as is the habit of Science and Sense, this column will endeavor to provide a common-sense approach to Being Prepared.

Most people have a pre-conceived notion of what it means to "be prepared." Of course, everyone's interpretation is different, sometimes wildly so. Some people "prepare" for the 24-hour power outage, a fairly common occurrence. Others are somewhere in the middle, squirreling away food, water, and other supplies to last weeks or longer. Then there is the hard-core prepper, prepared for months to years of self-sufficiency. For the purposes of this discussion, I will use three categories – 1) The Casual Prepper, 2) The Serious Prepper, and 3) The Survivalist. Each category uses a mental picture of the likely scenario for which they endeavor to be prepared.

As I write this, the song "Don't Worry, Be Happy" just came on the radio. This is a good description of the Casual Prepper. Mr. Casual likely has some cans of soup, other canned food, and maybe a 2 to 5 gallon supply of drinking water ready. He also has candles and flashlights with fresh (ish) batteries.

Mr. Casual is prepared for the 24 to 48 hour power outage. I don't know about you, but my house, in the County but on the fringes of developed Prescott, has lost power several times in the last five years, usually for a few hours, but a couple of times, one to two days. The Casual Prepper approach is generally fine for this kind of common disruption of services. And in none of those power outages did we lose water or gas service. The Casual Prepper approach works for a basic event of inconvenience.

Think of the Serious Prepper as the person who expects a longer disruption, days-to- weeks. Mr. Serious is concerned about major disruptions that might come from a hurricane, a big forest fire, or some kind of societal uprising that would disrupt the chain of delivery. This author falls into this category. Mr. Serious has months of shelf-stable food storage, a method to collect and purify rainwater (I can store 725 gallons at a time), a generator with lots of fuel (mine is a dual-fuel unit, gasoline or propane), at least a cord of firewood (if you have a wood-burning fireplace), at least three months of medications and dog food, and (this one might "trigger" some people) plenty of ammo. The food, water, firewood, medications, and other items are obvious – you're gonna need to eat, drink, stay medicated, stay warm, and feed your furry friends. The ammo provokes a darker subject – Mr. Casual is going to get hungry, and go looking for supplies. Mr. Casual might just transform from Mr. Rogers into a savage zombie rather quickly. The ammo protects Mr. Serious' supplies, and also could serve as alternative currency if the monetary system crashes (likely).

The Survivalist assumes that the disruption will last for months or years. This guy is ready to go off-grid, live for months to years on stored food, fishing, hunting, and gathering. Think about the teenagers in "Red Dawn". The Survivalist, while prepared to dig in at home, also has a bug-out location, probably way up in the mountains, that has a secondary stash of goods, and possibly a defensible underground shelter. There is a booming industry in pre-manufactured bug-out shelters that can be built at a factory and delivered to a site for burial. You might be surprised how many of these secret sites are out there.

Of course, we don't know what the disrupting event might be. Short-term weather or fire issues are somewhat expected. It is the North Korean or Chinese EMP attack that is not predictable (that event would fry all of our electronics, and send us back to the 19th century for months to years). Also possible is the massive geomagnetic storm from our sun, which would wipe out satellites and power grids for weeks to months. And of course, we all saw how a bio-engineered viral weapon used fear to disrupt our lives five years ago. Lots of doomsday scenarios are out there, so I won't belabor the point - look up articles online, if you need more reasons to lay awake at night, worrying.

Wow, stocking a Prepper Closet sounds expensive, you say. Well, it can be, if you do it all at once. A good method is to just pick up a couple of extra items with every trip to the store.

Some canned corn or a case of bottled water is not going to break the bank, nor is an extra bag of dog food. Watch expiration dates so you don't waste food – we restock our Prepper Closet back-to-front, keeping the older stock out front for regular use. Also, you can buy pre-packaged meals with a shelf life of 25 years. We have some of these, and we've taste tested a few – they're not bad!

The bottom line here is that we live in a (sort-of) free country, and you can prepare for whatever scenario you like best. At the very least, everyone should have a week's worth of shelf-stable food, and enough water and meds to last a month. This way, when the event happens, we aren't all piling into Fry's in a frenzy of panic-buying. Of course, Fry's will run out of stock in just a few hours, so there's that.

Happy Prepping!

The Science of Getting Away

Lately this column has been covering some heavy topics – lots of science, community topics, and big words and thinking. Not today. As August rolls around, my mind turns to Getting Away. And I am not talking about a weekend trip, inevitably rushed to cram in as much recreating as possible into 48 hours. No, Getting Away means an actual trip, of at least a week, to someplace DIFFERENT. A chance to unplug, unwind, recharge, decompress, recompose, reset.

"Getting Away" means something different to each of us. Some people go to the same place every year – a cabin you might own, or a beloved lake or beach. Others mix it up, going to a new destination each time. Growing up, my family did a bit of both. We had a friend's lake house that we would visit a few times each summer for long weekends. Five hours from home, it was far enough, and different enough (Hey, water!) that it felt like a getaway. But the familiarity was there, and we could contrast the small changes from year to year.

My parents also felt it important to show the world to us kids (at least within driving distance), to illustrate that not everybody lived in a flat, ugly oil town. We took epic road trips – to Tennessee, Arizona, Utah, Colorado, and all over Texas (which can include a LOT of variety in landscapes). In the days before screens, my Mom would often read to us. Mark Twain was among her favorites, and our family lived by one of his quotes – "Travel is Fatal to Prejudice" (thanks MT). My Dad would often gaze out upon foreign landscapes like Tennessee, and wonder aloud – "There's no oil here, what do these people DO?" Another quote from a Colorado trip – "These mountains – too steep to plow, too rocky for oil - they're worthless". This one still rings in my head during my annual Colorado trip. My Dad was a true west Texan.

By the time this column runs, I will either be there, or be getting ready, for my annual high altitude off-roading trip. On a family trip when I was 5, we made an overnight stop in Ouray, Colorado – the "Switzerland of America". Being a flatland kid, I was awestruck by this gorgeous town in a box canyon, with waterfalls, hot springs, trees like I'd never seen, and oh my, the rocks! My career as a geologist was born on that trip, nurtured back home by being around petroleum geologists (my friends' Dads, and Dad's friends). On my annual trip, the trail days (up to 13,000+ feet) are a sensory overload of in-your-face geology, as well as 100-mile views, sudden weather

changes, and trails that often have a 500-foot drop, 9 inches off your tire track (no guardrails). The Ouray trip scratches several

226

of my itches in one week – geology, beauty, off-roading, hot springs (did I mention the awesome hot springs pool right in town?), weather, and most importantly, the feeling of being Away.

I make an effort to minimize screen time during this week. Up on the trails, there's not much cell service, so the phone is just a camera. In the evenings, after a hot springs soak, I let the outside world in through my phone, but just for a few minutes. The inn I stay at has a balcony over the Uncompahgre River in every room. Even though each room has cable TV, I keep the TV off, and let the sound of the river be my entertainment. Deer and elk walk along the banks, grazing, just off the balcony. Is Ouray Heaven? If not, it is definitely Heaven's next-door neighbor.

Not that this trip is my only recharge spot. My daughter lives on Ocean Beach in San Diego, and I do enjoy the beach in short stints. A couple of days of listening to the crashing surf is a tonic to my soul. I prefer the powdery-sand beaches and warmer/cleaner water of Florida beaches, but that is a bit too far to drive, and I am minimizing flying these days. I find White Sands National Park in southern New Mexico to be a calming place, with nothing but snow-white sand and blue sky to simplify the world for a while. Saguaro National Park and Park Link Road in southern AZ also are a balm to my soul.

Speaking of National Parks, I have visited 32, and I still have quite a list yet to see (plus revisiting favorites). National

Parks are one of America's best ideas, and they offer the full cornucopia of natural landscapes (North America is embarrassingly wealthy in natural beauty). State parks are also wonderful, and many are just as spectacular as National Parks. Fun fact, here in the Yavaplex, you are within a day's drive of 13-18 National Parks (depending on what a day's drive means to you), with a stunning variety of landscapes. Mountains, canyons, deserts, lakes, ocean beaches, cultural sites, geological wonders, all a short drive Away.

Of course, if you are pressed for time, and can unwind in less time, the Bradshaw Mountains and Prescott National Forest are in our backyard. Sedona's Red Rocks and the Grand Canyon are nearby, and these beauties attract visitors from literally all over the world! And we can do it in a weekend from here!

I urge you to prioritize Getting Away. The benefits are many, as are the opportunities. Some people say "he who dies with the most toys wins". My mantra is "He who dies after seeing the most natural beauty wins". I'm not done yet, but I've already won.

Getting Away – Part 2

After my last column about Getting Away, I've had several people ask for recommendations for National Park trips. Asking for my favorites is like asking which of my rock collection is my favorite – impossible to answer. So, I will provide four loop trips that can expose you to multiple park experiences, all starting out right here in the Yavaplex. Each one is a 7-10 day trip, but of course you can add or subtract from it to fit your time, budget, and personal tastes in natural beauty exposures. A glaring omission is that I have not included the Grand Canyon, Petrified Forest, or Painted Desert. These are so close that I assume you have all visited, and if not, get after it! These four loops take you southeast, northeast, north, and west from here, and each one offers variety. All four can be done as camping, RVing, or hotel-stay type trips. The first two are trips best taken in any season but summer, and the last two are best taken in any season but winter.

Loop #1 – Deserts, sand, and caves. This one takes you to some of the parks of my childhood, so I have a soft spot. Start out by going south to Tucson, and take in Saguaro National Park. Tucson offers lots of great resort hotels, so you can pamper yourself the first night or two. From there, divert through Tombstone and overnight in Bisbee – not National Parks, but worth the visit on the way. Go through Chiricahua National

229

Monument, then spend a day at White Sands National Park in New Mexico, and stay overnight at 9300 feet at The Lodge at Cloudcroft. Next day, drive to Carlsbad Caverns, and if you have time, go over to the Alien Museum at Roswell (very cool). Next up, Guadalupe Mountains National Park, and hike to the highest point in Texas! From there, take the big trek to Big Bend National Park. Magnificent desolation, amazing river-cut canyons, and incredible geology. Come back through Faywood Hot Springs near Deming, NM and soak away the miles.

Loop #2 – Deserts, Mountains, Big Trees, and Weird Trees in California. This trip starts by going to Las Vegas, and doing a side trip to Valley of Fires State Park. Not a NP, but it should be – it's a geologist's dream. And you can do it in an afternoon. Say goodbye to bright lights and big city, and head to Death Valley NP for heat and desert vistas (stay at the park's hotel, a true oasis). From there, a total change in landscape is Sequoia and King's Canyon NPs, not too far away. Yosemite is next up, many peoples' favorite (and the crowds prove it). Then a long drive south to Joshua Tree NP (Joshua Trees look like something from a Dr. Seuss book), with interesting geology. If you're feeling dry after this loop, you can divert to LA, and take a boat ride to Channel Islands NP, a unique biome for sure. Then get the heck out of LA and come home.

Loop #3 – Canyons, Crazy Rocks, and Arches in Utah. Start by going up and around the Grand Canyon through Page, stopping at the Lake Powell Dam visitor center. By nightfall you will arrive at Zion National Park Lodge in Springdale, Utah. This is the beginning of your eastward trek across Utah's "Big Five" National Parks – Zion, Bryce Canyon, Capitol Reef, Canyonlands, and Arches. Although it may seem like these five parks are all just interesting rocks, don't be fooled. Each park has its own character, and lush valleys offer a counterpoint to the sometimes insane rock formations. I like this trip best in fall (October or November), because the light in the canyons is magical for photography. End up in Moab, and take some offroad trails on a bike or in a vehicle, whichever suits your fancy. Moab is Mecca for this activity.

Loop #4 – Serious Mountains, Hot Springs, and Culture in Colorado. This is the trip that I do parts of every August, and have for my whole life. Start out by driving to Durango, and take the Million Dollar Highway to Ouray. Bask in the "Switzerland of America" with its high altitude trails and hot springs. Then a short drive northeast to Black Canyon of the Gunnison NP – almost as deep as the Grand Canyon, but in some places, the rims are close enough that you can holler to hikers on the other side. Then north to the Grand-daddy of hot springs, Glenwood Hot Springs. Continue northeast to Rocky Mountain National Park, a showcase for the "Roof of America", with numerous 13,000 and 14,000+ ft peaks, and an amazing abundance of wildlife. Wind back to the south to Great Sand Dunes NP, A little bit of the Sahara on the foothills of the Rockies - unexpected and amazing. Hit a couple more hot springs next (Joyful

Journey, Cottonwood, or Pagosa Springs), and circle back to Durango. Finish off the trip with Mesa Verde NP, for a window into how people lived over a thousand years ago. If you have anything left, you can pick up Petrified Forest and Painted Desert NPs on the way home.

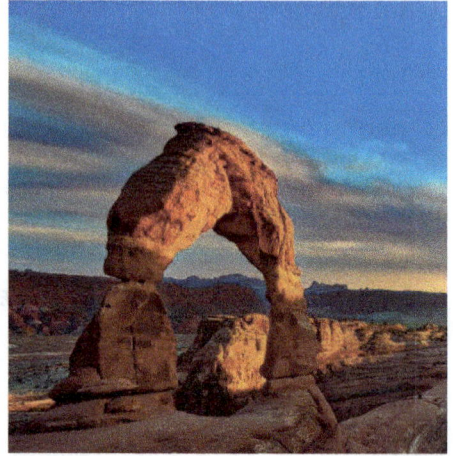

Whew! I'm exhausted! Loop trips are a big bite, and not for everyone. So, you can break off pieces of these loops and do one or two parks at a time, and really immerse yourself for a few days. Either way, I urge you to get out there, and enjoy these amazing places.

As Americans, you OWN these parks. It would be a shame to not see them while you can.

Happy Parking!

The Science of Fiscal Sanity and Haircuts

It was a sad day in October 2024. I dropped and broke my trusty Wahl clippers on a rock in the back yard, ending a 32 year period of service for these amazingly well built clippers. Okay, that lead-in requires some explanation, as does the title of this column.

Since the mid-1980s, I have sported a short flat-top hairstyle. It is worth noting that if I let it grow, it becomes a thick, curly shag carpet that holds in heat like an old-school Thermos bottle of coffee. The short flat top eliminates the shag carpet problem, and in my opinion, looks good on my blockish cranium. It also keeps me from having to spend any time at all on daily hair grooming, and has kept me from owning a blow dryer since the first Reagan administration. Up until 1992 (half my life!), I visited a barber every two weeks, spending the exorbitant amount (at that time) of $11 to get a five minute job with the clippers. Also, our first child was born in 1992, I was looking for ways to "trim" our budget to afford diapers.

Enter Sally Beauty Supply and their amazing Wahl Clipper set! I sprung for the Professional model, just like my barber used, and it was $44 with tax. I figured at that cost, I would break even in about four haircuts (two months), likely the period of time that it would take me to perfect my technique. So, I began my every-two-week ritual of self-haircuts. They say that it is a fool who serves as his own attorney, surgeon, or barber, but I threw caution to the wind. The nice thing about short haircuts is that any mayhem grows out and covers the mistake quickly.

The process has been the same since Bush 1 left office. I go full-hillbilly, and bungee a mirror to a pole in the back yard. A short extension cord and the magic Wahl clippers, and I am in business. Early on, I did make a few errors, so for a few months I went with just a buzz cut, no flat-top shaping. But once I got braver, it only took me a couple of cycles to get that Kansas-flat, shiny-scalp-in-the-middle flat top look that I craved. Shaping the rim was a challenge due to some stubborn cowlicks, but I eventually got the hang of it. And for all these years, this has been the procedure. I have been blessed with hair like Stalin, so baldness has never been an issue. I also became the neighborhood barber to my son and his friends, up until they were about 12 and discovered girls. Buzz cuts apparently don't attract modern young ladies.

Style and process aside, my main point here is the fiscal advantage of doing my own 'do. I have a penchant for classic cars, and I have owned one even during lean times when the kids were small. Sweetie would occasionally frown at some small purchase to keep the classic running, so I devised an (admittedly) self serving justification – the Wahl clippers paid for the cool old convertible. Sweetie originally met my story with skepticism, but being a solid, logical sort who is blessedly open-minded, she eventually came around to my way of thinking. And I now have a classic car, a super-stylin' haircut, and a happy wife.

So, for any of you fellas out there who wish to walk the path that I have paved with my flat-top – here is the math. A haircut every two weeks, for 32.5 years (that's around 400 months, at two haircuts per month, plus occasional trims – so let's say 860 haircuts since March of 1992 (I am not correcting for leap years, five-week months, or the rotation of the Earth). The going rate these days is $35 for a standard haircut (gulp), and it was $9 when I started, so let's use an average of $26 per haircut for the 32.5 year period (I'm not counting tips, gas back and forth, or time lost waiting). Crunching the numbers, my grand total saved is $22,360, less the $44 for the clippers in 1992, so $22,316 net.

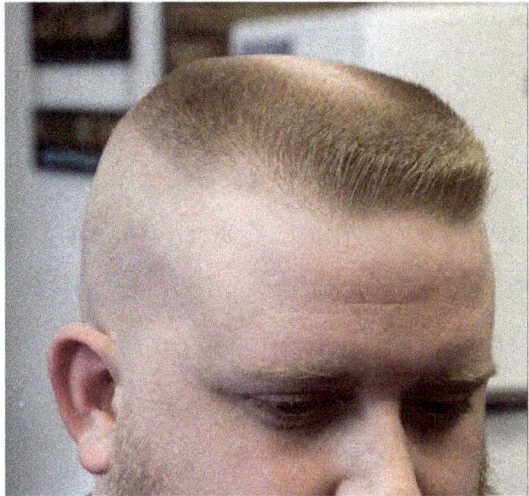

Now, I don't want to disclose what I have spent on my classic rides, but that amount of money justifies a big chunk of the cost. I have also saved classic-car money over the years as a borderline cheapskate, mostly in clothing myself, taking my lunch when I had an office job, and doing most of my own yard work. Yes, I am of a generation that bought extras using cash, and if I needed more money, I worked for it. Keeping the credit card balances low (or zero) is a superpower of my generation.

I made this century's trip to Sally Beauty Supply, to buy another set of Wahl Professional clippers. I was thrilled to see that they offer the exact same unit, and it has always been Made in America! Not surprisingly, they had gone up in price, this time $146. The new, higher price still pencils out to break-even at four haircuts, given today's rates. Exactly the same as in 1992! So my Return on Investment should continue for the rest of my life. I have been pondering what to do with my old, broken, trusty set. Rather than unceremoniously tossing them in the trash, I think I will tuck them into a corner of the trunk of my classic car. Those clippers certainly earned that ride.

The Science of Being an Opinion Columnist

Having written dozens of columns so far (including many that haven't run yet), I have come to some conclusions about this endeavor. With a goal of providing explanation and opinion about topics with a scientific element, I use a variety of investigative sources, plus my personal experiences, plus common sense (which really isn't that common anymore). This recipe provides a truth- and fact-based argument that I can stand behind. Yet rebuttals fly, often rooted in incorrect conclusions drawn by people using biased sources, little experience, and an utter lack of common (or uncommon) sense.

A major drawback to this form of media (print and published-online content) is that I make my point once, yet rebuttals can come in for weeks, and by the dozens. My Editor does a pretty good job of sifting through the Rants (and Raves, the paper has a "Rants and Raves" section) and printing what he feels are the best ones, but the numbers game is stacked against the columnist. For example, my five part series on Energy ran last March. My column about Roundabouts ran in May. My three-part Climate series ran in June. Months have passed, and people still write rebuttals to these topics, some well written and intentioned, others scurrilous and filled with inaccuracies. A few have tried to goad me into a back-and-forth, but that is not productive, and frankly would get boring fast. That kind of "neener neener" is for social media, and grade school playgrounds (not much difference there).

Another interesting facet of reader feedback, is that in person and by direct email or text, I get about 95% support and concurrence. When I asked my Editor about this inconsistency, he replied that it is common to get a higher rate of negative comments to the paper than positive, and a higher ratio of positive to negative when the comment goes directly to the author. This is a function of human nature. When people disagree or are angry, they tend to write in, often anonymously. This provides a way to "fire back" from behind a wall. Direct responses to the author leave the commenter unprotected, without the wall of anonymity. Therefore the direct comments trend toward the positive. Sometimes people write in an anonymous "Rave", which runs against the trend. These are probably folks (like me) who still use written thank-you notes in certain societal situations.

Social media has also trained many people to think that they must weigh in on every little thing. In print media, there is no immediate "reply" function. There is the anonymous "Rant", which allows responders to be anonymously mean, just like on social media. Another factor is that responders of a "certain bent" are just nasty. They have been taught this tactic by their ideology. Those of us who tend toward civility often recoil from this aggressive nastiness, and just don't engage (which is what the nasty responder wants, to shout down the civil person, and "win" by sheer volume and vitriol). I recognize that personal attacks and general animus has been ginned up by social media, and that people act differently face to face (even people who vehemently disagree with me), than they do from behind the wall of anonymity.

I have also been accused of name-calling ("greenies", "lefties", etc.) These terms I have indeed used, but only in columns that cover a topic where the other side has used far more vulgar and vitriolic language. And "greenies" identifies those with a "green" agenda. "Lefties" identifies those whose worldview is that of leftism. These aren't epithets, just descriptive terms. Believe me, if I wanted to, I could use far more acidic language, but I keep it mild just so the opposition can get a tiny taste of the nastiness that they serve in heaping helpings every day. And my parents taught me not to be mean, just honest.

For those who regularly write in with just plain hatred for anything I write, I hope you feel better. My Editor has done a great job of keeping things civil in the Courier. For those of you who enjoy my columns in silence, I thank you. For those of you who don't enjoy my column, I ask you – why do you read it? Isn't there enough aggravation in this world already?

Oh, and I don't plan to stop. I have about 45 more topics in the queue, with people bringing me more ideas every week (keep 'em coming!). I focus on the positive responses, and just accept that some people need to vent their bile. As I said before, I hope the venting makes you feel better.

One final suggestion, whenever someone references one of my columns in a rebuttal, I urge you to use the search function, and go back and read my original column on the topic. So far, most of the issues being rebutted were addressed in my original column, and in many cases, the "rebuttal" helps me make my points. I thank you for that.

And the number of responses, both positive and negative, tell me that my main goal is being met. I am inviting conversation, debate, and discussion – whether it is over coffee, over your fence, or over the phone. Getting folks to think, and talk amongst themselves about topics of the day, IS the payoff for me.

So, thanks for reading, lovers and haters alike!

EPILOGUE

I mentioned Marvin Longabaugh in the Introduction to this book. Marvin re-lit my writer's pilot light in 2016, by asking me to write Opinion / Editorial columns for his small town paper in Texas, the Navasota Star. Marvin was a human hurricane, a force of nature (for good), and the kind of character that Midland, Texas produces. I consider myself lucky to have grown up around such people, including bright stars like Marvin.

Science and Sense would not have been possible without Marvin's persistence and goading for me to start writing again in 2016. When he passed away in March of 2017, it was mercifully sudden, as he was preparing to judge a chili cookoff that he had started in Navasota. I was out of country when I heard the news (same day – our high school community is well-connected). Knowing that the paper would likely fold, and I would have one shot to write something about Marvin for the final edition, I sat down that day and wrote the following piece for the Star. It flowed out of my fingers and into the computer in one long flow of emotion and gratitude for having known such a person.

So, readers, allow me to introduce you to Marvin Longabaugh.

Contemplating a Life Well Lived

Marvin Longabaugh was larger than life.

I know that my friend would relish the obvious pun about his girth, but the reality is that Marvin got it right. Big hearted, brilliant, and giving, here is a guy who I am sure was welcomed into Heaven with a hearty "Well done, sir!" from St. Peter.

Most of us are likely past the awful initial shock of learning that he has gone. Next up in the grieving process is coming to grips with what that person meant to us. So, in my scientist's mind, I have been trying to glean why Marvin was so beloved.

Since my regular article in the Star is "Scientifically Speaking", and I am an analyst by nature, my goal is to try to convey what made Marvin Marvin, so that maybe we can all bring some of his magic into our lives.

First, one must understand where he came from. Those of us who grew up in Midland know that it can be a demanding place. The landscape is bland, the dust storms and summer heat are brutal, and the cold winter wind blows straight from the Arctic (nothing to stop it). As a one-industry town (oil), local economics and personal financial stability are never a given. Midland also has a high through-put of strangers – newcomers seeking their fortunes in the oilpatch. So, in order to survive,

Midland requires toughness, persistence, and great people skills.

Midland is also filled with optimists – people who endure and survive the hardships with a smile on their faces (sometimes filled with grit from a dust storm). Marvin and I grew up around dreamers, hard workers, and characters, in a much higher concentration than most places (and that is saying something in Texas). Marvin once said that Midland is more Texas than Texas. Wise man. Midland is a crucible for character development, and was Marvin a character, or what?

Second, Marvin grew up in a home where Dad was a Naval Officer, an unusual avocation in landlocked Midland. Marvin and his brother Keith were often the "men of the house". This must have honed Marvin's sense of service and responsibility.

Third, Marvin was a fat kid. So was I. We shared a learned skill of a quick wit, and learned that humor was an excellent defense against larger, stronger tormentors. Inasmuch as I suspect that Marvin's size contributed greatly to his quick wit, I am certain that it was simply baked into his DNA. Had his container been that of our handsome quarterback Doug Atnipp, I'll bet Marvin would have still been just as warm, self-deprecating, and funny.

Fourth, Marvin and I benefitted from the incredible quality of our peers. The people we grew up with went on to do amazing things in their lives, and I am not just talking proud here. Truly gifted, accomplished, larger-than-life people sprung from our ranks. There must have been something in that hard Midland water in 1959-60 that did this to us – a class of Supermen and Wonder Women.

Beyond our years in Midland, we grew apart, as often happens. Marvin gained his law degree from Texas Tech, made a mark in the Texas Jaycees, then spent most of his career in Las Vegas. He worked in family law, and later helped homeowners to keep their homes during the real estate crash of 2008-2010. In the early 2010s, he "retired" back to the life of a Texas country lawyer in Navasota, with his beloved wife TJ by his side. His initial idea of living a slow-paced life left him with time on his hands. He chose to channel that energy into something positive, and the Navasota Star was born. That little paper, swimming against the tide of digital media, proved that hard work, heart, and community-mindedness still exists, and can flourish.

So, what of Marvin's impact, and my exhaustive scientific analysis? The scientist in me knows that each human is an individual experiment, a mix of unique ingredients that cannot be replicated. Marvin was truly unique, and in this scientist's opinion, embodied the best of us all. Physically he is gone, but I know that his humor, spirit, humanity, and dedication to improving the world around him will live on in each of our hearts. It was not possible to know Marvin, and not be changed for the better by the experience. As a lifelong friend of Marvin's I can say, with certainty, that I want to be him when I grow up.

This picture of Marvin has a story. He was trying to be good on his diet, and he knew that receipts were the paper trail. He learned that if he ordered a salad, 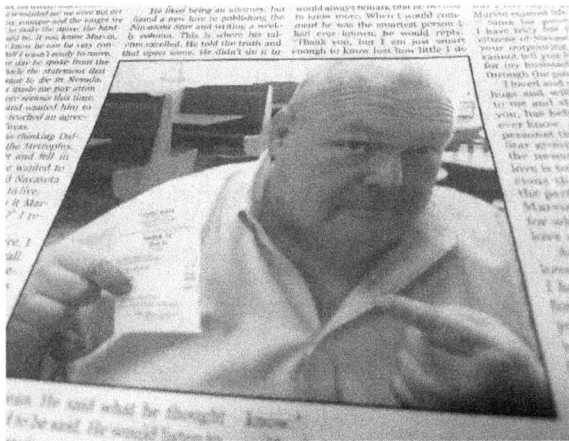 that was what was on the receipt. He also learned that substitutions were not captured on the receipt. Therefore, he substituted a hamburger patty for the lettuce, a slice of cheese for the tomatoes, and a bun for the salad dressing. Voila, a cheeseburger, with the paper trail of a salad. Marvin was the master of his universe, a lawyer who knew how to spin a yarn. And how to cover his dietary tracks.

Author Bio

Growing up in Midland, Texas, during the 1960s and 70s, Kelly Kading experienced the dynamic world of an oilfield town. Surrounded by dreamers and hard workers, he developed resilience and curiosity amidst the boom and bust cycles. His mother instilled in him the values of learning and logic (and how to make a mean pecan pie), while his father taught him hard work and integrity (and how to catch the big bass).

Kading's educational path began in Midland Public Schools, continued at Texas Tech, and led to a Geology program at Colorado State University. Initially entering oil exploration in Dallas, the mid-1980s crash shifted his trajectory. After a stint owning a ski shop in Colorado, he transitioned to "Environmental Geology," working in groundwater cleanup for clients such as Texaco and Mobil in south Florida.

His career took him across the country, eventually settling in Arizona. Now, Kading consults on environmental projects, explores the American Southwest's natural beauty, and visits National Parks (32 and counting!).

Today, Kading manages a small consultancy focused on environmental assessment and cleanup projects, primarily within Arizona's infrastructure sector.